INTERPLAY
INTERACTIVE
DESIGN

AUTHOR BIOGRAPHY

Lauren Parker is Curator of Contemporary Programmes at the V&A Museum, specializing in new media, audio and interdisciplinary projects. She has curated several exhibitions and events programmes, including the V&A's first ever audio exhibition, Shhh... (V&A, 2004).

ACKNOWLEDGEMENTS

My grateful thanks are due to all the designers, programmers, artists, manufacturers and retailers featured in this book for their generous cooperation, and for the supply of photographs and related design material. In particular I would like to thank the following for their support of the project: Matt Adams, Anthony Burrill, Bob Greenberg, Takeshi Hamada, Antti Hinkula, Richard Holley, Rei Inamoto, Alexandra Jugovic, Nathan Lauder, David Rainbird, Florian Schmitt, Bob Shevlin, Daljit Singh, Karen Spiegel, Teemu Suviala, Sam Taylor and Catherine Williams. I would like to thank Jane Pavitt for her encouragement and good advice, V&A Publications, in particular Mary Butler, Clare Davis and Monica Woods, for their support of this project, Krystyna Mayer for editing the text, and Graphic Thought Facility for designing the book and series. Thanks also to Ann Gelbmann, Abigail Billinge and Frances Ambler for picture research, and to colleagues at the V&A who contributed advice and suggestions, especially Shaun Cole, Susan McCormack, Rebecca Milner, Guillaume Olive and Carol Tulloch. Final thanks go to Jan for his encouragement and support.

SERIES FOREWORD

Design is an essential component of everyday life, in ways that are both apparent and imperceptible. But who are the authors of the things that surround us? What drives the thinking behind the development of new products? From cutting-edge experimentation to the re-styling of mass market goods, designers leave their imprint in a diversity of ways. They shape the objects with which we furnish our homes, the tools with which we communicate and the environments in which we live, work and play.

The V&A has played a role at the heart of design culture since it was founded in 1852. Part of its mission was a commitment to the understanding of contemporary design, and this mission remains true today. The V&A now embraces all aspects of design in contemporary life, from fashion, graphics, photography and digital media to craft, architecture and product design.

The V&A Contemporary series explores the designer's role in shaping products of all kinds – from the one-off to the mass produced, from objects in three dimensions to digital environments. The series celebrates creativity and diversity in design now, highlighting key debates and practices, and confronting us with questions about the future of our designed world. Each title takes a critical and informed look at a particular field, built around interviews with designers and commentaries on selected products and projects. We hope these studies encourage us all – designers and consumers alike – to look with fresh insight at the objects and images around us.

Jane Pavitt, Series Editor

6-39
SCREEN VISIONS
COMPUTER TECHNOLOGY APPEARS IN ALL AREAS OF OUR DAILY LIVES.

Computer technology appears in all areas of our daily lives – from mobile phones to microwave ovens – at home, in the office and in the intangible networks of telecommunication, entertainment and information services that surround us. When machines were mechanical, there was a direct, physical way to interact with them: you flipped a switch or turned a dial. The everyday relationships we have with computer-based technology are different. They require interactions that are more complex, less transparent and increasingly intangible. We rely on these digital devices, and yet our interactions with them can be awkward, lacking in pleasure or downright confusing.

Interactive design is an area of study aimed at understanding how we relate to technology, and developing new design methodologies to create products and services that merge aesthetics and culture, technology and the humanities. For Irene McAra-McWilliam, head of the Interaction Design department at the Royal College of Art in London, interactive design is a discipline that 'explores the relationship between people and technology, [occurring] in the space where users and technologies meet, and interaction designers inhabit this space, looking in both directions simultaneously'. [1] This notion of interaction embraces both an external perspective, the aesthetics of design, and an internal awareness – of form, content and the architecture of information.

Designers, academics and technologists are seeking to discover how the use of these new networked technologies will impact on society and how the changing world will create requirements for new design skills and new design visions. Interactive technologies are creating a need for a new kind of design – a fusion of graphic, architectural and product design with an awareness of time-based narrative and navigational structures.

At the heart of interactive design lies an understanding of networked media – in particular the Internet, but also handheld communication devices and networked physical environments. 'Interplay' presents a series of case studies with designers and their clients that examine the creative solutions, imaginative use of technology and unique user experiences created within these networked spaces.

WWW.WDYWTDWI.COM
Design Agency: Fibre
Creative Direction and Design: David Rainbird, Nathan Usmar Lauder
Design: Tommy Miller
Game Programmer: James Robertson
Client: Motorola/The Fish Can Sing/ICA
2002

PHOTOS

Design is context dependent – it links to the society that surrounds it. To consider design without an awareness of the cultural context is to consider it as an independent craft or art form, not as a form of communication. At the same time, designers are facing new challenges: to adopt a range of roles – from artist to collaborator to storyteller and mythmaker, to embrace a multidisciplinary stance when creating personal visions of the world around them and to broker new kinds of relationships between clients and users. At the heart of these new challenges lies the need to draw together wider cultural and social trends with future advances in technology. Within these intertwining strands, designers are creating a new vision of how individuals can discover more emotional, experiential and tactile relationships in the all-encompassing digital environment.

Designing for interactive media is in its infancy. The earliest examples stem from the birth of the computer game industry in the 1960s; however, the web design industry is much younger, celebrating the tenth anniversary of the World Wide Web in 2003. Mainstream recognition of web design arrived in 1997 with the introduction of an interactive category to the British Design and Art Direction awards (D&AD), celebrating the work of interactive designers alongside other areas of design for the first time in the UK.

The digital design environment at the beginning of the twenty-first century has become a mixed marketplace, with opportunities for freelance designers, loosely formed collectives, boutique agencies and larger-scale multidisciplinary consultancies. For many younger, smaller design agencies, interactive design forms only one part of a wider collaborative strategy. For David Rainbird and Nathan Lauder, co-founders of London-based agency Fibre, their cross-disciplinary stance has evolved in part due to the rapid technological advances of the past decade: '…[we] graduated in 1990 – right at the start of the computing boom. This meant that in our working lives we have worked through new opportunities – 1995 and the rise of the CD-ROM, the boom in the Internet between 1996 and 1997, and since then the video desktop revolution. This has meant that we have ended up being multidisciplinary designers.'

Since the late 1990s there has been a growing recognition that interactive design is a key constituent of graphic design practice. Designing for interactive media (in particular web design, but also CD-ROMs, in-store kiosks and other examples of interface design) has become a core practice for many design groups. In June 2003, the UK-based magazine 'Design Week' published its annual list of the top design groups. Of the top fifty companies, thirty included designing for digital media as part of their remit. Designing during this digital age creates new imperatives and new challenges; however, the fundamental structures and concerns for web designers remain inextricably linked with the histories of graphic design, moving image and architectural practice.

SCREEN VISIONS: RE-IMAGINING THE WEB

Designing for the web has at least three unique elements to it. Firstly, the web can be like software, allowing the user to engage with interactive and non-linear elements. Unlike turning the pages of a book or magazine or flipping channels on a television set (where the viewer changes between one static, predetermined presentation to another), interactive media encourages the user to change the content itself, playing a major role in the outcome of the site. Secondly, the web synthesizes different media elements, combining film, animation, images, text and sound. Finally, because the web is a real-time worldwide network, it allows a level of collaboration and improvisation between multiple groups of people independent of geographical location in a way that is not possible in other media. Because of these time-sensitive and three-dimensional qualities, the web cannot be considered solely as an element of graphic design – it contains and reflects elements of architectural practice and even product design.

It is the nature of the web as a constantly changing environment that excites Japanese designer Yugo Nakamura: 'I think one of the key features of the web environment is that it is in a constant state of flux; it evolves from moment to moment, according to the user's intentional browsing activities, and the accidental discoveries she makes on the way… what I want to create is a unique communication experience.' [2]

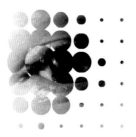

WWW.SURFACE.YUGOP.COM
Designer: Yugo Nakamura
2001–

WWW.SURFACE.YUGOP.COM
Designer: Yugo Nakamura
2001–

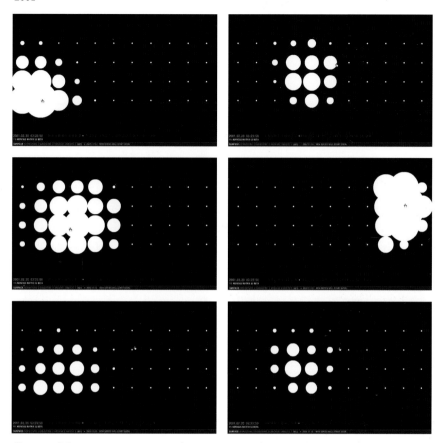

Some of the most exciting and creative work carried out today on the web comes from designers who are aware of the unique qualities of the medium, and are able to shape a new design and communication paradigm to embrace them. The peculiar characteristics of the online environment, for example, include technical considerations – from cross-platform compatibility between different browsers and operating systems, to the difficulties in ensuring a consistent user experience in an environment where the viewer is able to control the browser window size, colour calibration and screen resolution of their own computer. Most crucially, the dial-up speed that is available to the user can have a fundamental impact on the experience they receive. The rapid growth in broadband is offering greater freedom to designers to experiment with faster download speeds, more dynamic content and streaming media – from live music to video feeds.

WWW.DELAWARE.GR.JP
Exhibition JPG, Barcelona
Designer: Delaware
2002
(see also pp. 14–15)

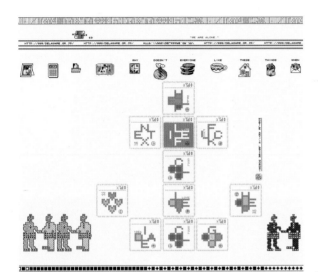

Other more experiential challenges for the web designer include the use of colour online. Unlike the light-reflecting and absorbing quality of pigment on paper, colour on the computer screen is illuminated. Light shines through the colour. London-based designer Anthony Burrill feels this allows him the freedom to create a different kind of experience from that available through the medium of print: 'I don't really miss the tangible nature of print when I am working on the web. You can do things with colour and with movement on the web that are amazing. I like bridging between things.'

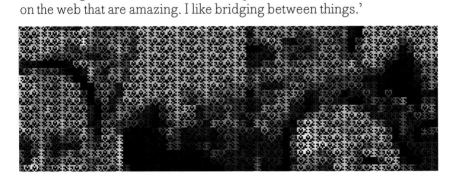

For designers who are interested in unpicking the underlying nature of the
online environment, two key elements lie at the heart of the design process.
The peculiar nature of the screen interface as opposed to print-based
media creates a different kind of relationship between the viewer and the
work, and this tension between on-screen and off-screen has informed
the work of a range of designers. In addition, designers have embraced the
interactive qualities of the online experience, encouraging continuous
dialogues between content and viewer, 'author' and 'reader'.

The nature of the screen interface, with predetermined screen size and
browser window, limits the available page scale. For designer and illustrator Karen Ingram, this means making choices about the content of her
work: 'Scale is a big factor for me as far as subjects go – in print, I'd be
working larger. My subject matter has deviated from what I typically
choose for painting or print – I find myself being more attracted to smaller
objects in my [online] work, to make my work more efficient as animated
objects. In print or painting I did a lot of work with the human form, but on
the web, I simply feel as though I cannot get the physical space to play
around with the human form, except in bits and pieces.' [3]

Designers are experimenting with ways of negotiating these limitations in
size and scale. Japanese design group Delaware have avoided the restrictions of the computer screen by playing with the conventional use of the
browser window and site navigation. Delaware replace the usual hyperlinks and vertical scrolling of a typical website front page with a linear site
created through continuous horizontal scrolling.

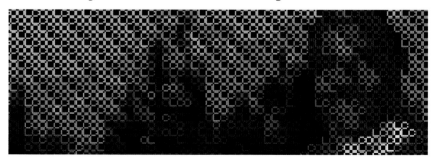

YOU'LL BE SOON BORN.
君ハ モウスグ 生マレル。

WE ARE HUMAN BEING

僕ハ モウスグ 死ス。
I'LL BE SOON DEAD.

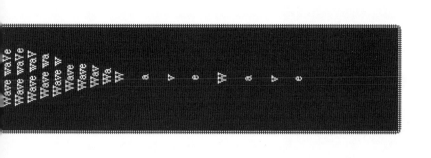

WE·LL MEET AGAIN·

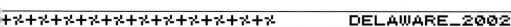

WWW.FROSTDESIGN.CO.UK
Design Agency: de-construct
Creative Director: Fred Flade
Design Director: Vince Frost
Technical Director: Matthew Wright
Client: Frost Design
2003

UK design group de-construct have devised different solutions to avoid the limitations of the browser window. De-construct have created a front page for the **FROST DESIGN** website that operates as a visual overview of the entire site, divided into a grid system that expands and contracts to allow the user to see additional text and links or to navigate the front page on a magnified scale. For their own site, de-construct have taken a more radical approach to the visual presentation of information – rejecting a graphics-based user interface in favour of a largely text-based search engine system.

The computer screen acts as a distancing mechanism between the design and the user. The communication that occurs between viewer and designer exists in a visually mediated form, without the possibility of any tactile or tangible connection. Beyond the computer screen users become disembodied; the body disappears only to be replaced by an arrow cursor, data glove or mechanical arm, while the individualized self is replaced by a computer software program or a set of rules, commands and functions. The screen serves multiple functions – acting as a view finder, camera, projection surface or semi-permeable interactive interface – however, each of these roles ensures that the user maintains a more or less distanced relationship with the content beyond the screen.

WWW.PROTOKID.COM
Design Agency: Bloc
Designers: Rick Palmer, John Denton, Eleanor Wilson
Programmers: Steve Hayes, Eleanor Wilson,
Luis Martinez, Jan Fex
Client: Diesel Kids S.rl
2003–4

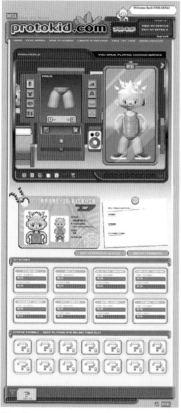

PUPPETTOOL
Designers: Frédéric Durieu,
Kristine Malden
LeCielEstBleu
2001

Designers have taken a number of different approaches to try and break through the barrier that the screen interface creates. These have included the creation of avatars (online representational characters) on sites such as Diesel's **PROTOKID** world, and immersive online narratives, such as the **LEXUS MINORITY REPORT EXPERIENCE**, as well as attempts to recreate haptic experiences beyond the screen. Designers Joshua Davis and Yugo Nakamura, for example, have both experimented on their personal websites with ways of recreating sensations of touch, weight or resistance through online objects and menus that simulate the effects of gravity and friction. The **PUPPETTOOL** program created by designers LeCielEstBleu encourages visitors to interact with animations or to create their own which can be uploaded onto the site.

CALENDAR
Calendar #4 Hanabi
Designer: John Maeda
Client: Shiseido Co., Ltd
1997

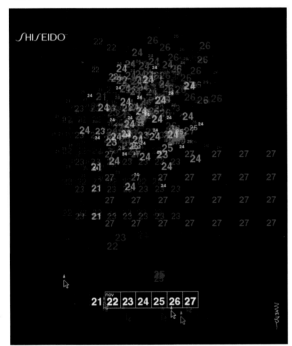

Japanese designer and author John Maeda is one of the most influential champions of interactive design working today. His design vision centres on the expression of the essential nature of computing and code. At the heart of his work lies the notion that the computer is not merely a tool, but a powerful means of expression. Interested in defining a new paradigm for the computer that rejects existing physical analogues – the printed page, desktop or cinema screen – to define the nature of interactive media, Maeda instead has chosen to explore the '…infinitely multidimensional space of pure computation', producing work that sits artfully between abstraction and craftsmanship. [4] Yet he takes a critical approach to the role of technology and is aware of the intertwining relationships between design, technology and society. For Maeda: 'technology is not a fad; it is the curse of our society. The question is, how can we control this evolution?'. [5]

GRAPHICS FOR SAWAYA & MORONI (DETAIL)
Graphic for Table 01
Designer: John Maeda
Client: SAWAYA & MORONI
2001

Maeda's strengths lie in his awareness of society's desire for emotion and intimacy, of the need to invest technology with 'humanist' sensibilities. A recent project, **OPENATELIER**, is an attempt by Maeda to create freely accessible digital art tools and includes the creation of the **TREEHOUSE STUDIO**, a research project to introduce children and adults to the idea of digital art and design through a series of tools, exercises and online activities.

TECHNOLOGICAL VISIONS: INNOVATION SHAPING STYLE

The short history of web design has been driven by a series of rapid technological innovations – from the launch of the Mosaic browser in 1993, which heralded the birth of the Internet as a commercial force, and the development of Java technology in 1995, allowing interaction between user and server for the first time, to the release of Macromedia's Director and Flash programs, to 3D modelling software and beyond.

This focus on a continual technologically driven innovative momentum sits within a wider new media critical framework. Since before the existence of the web, cultural commentators have imagined utopian virtual worlds shaped by technological progressivism – a continual advancement in technology and innovation. Commentators including Timothy Leary, Jaron Lanier, William Gibson, Howard Rheingold and Nicholas Negroponte conceived virtual space as a place of transcendence, a neutral place existing both as 'elsewhere' and 'nowhere': 'the virtual world is a place that is absolutely different. The "otherness" of this virtual space is now conceived as the ultimate utopian destination.' [6] This critical framework presents technological progressivism as a distinctive social vision: seeking to persuade us of the utopian qualities of mediated communication, electronic communities and virtual (post-) selfhood. It is within this wider critical context that interactive designers continue to work today.

Technology is not neutral: the flip side of these utopian visions includes a danger of a stylistic homogeneity, and the risk of using the available technology as a benchmark for adopting a ready-made style, rather than embracing the software as a tool for individual expression. Web design's evolution will continue to be linked to the development of new tools; however, this desire for technical expertise needs to be balanced with a confidence of creative expression.

For Matt Owens, designer at New York agency One9ine and creator of personal site **WWW.VOLUMEONE.COM**, 'technological proficiency is really applauded in the new media space... people love it, the technological is really revered. The visual is revered too, but in a different way. You have to hit the right balance – do something that's tricky by all means, but not for the sake of it... To me, it's more interesting to challenge yourself to be technical and visual: people tend to fall into one or the other. I think the space in between is really interesting.'[7] For web designers working today, this balance between technological proficiency and the presentation of personal creative expressions lies at the heart of the design process.

PERSONAL VISIONS: SELF-EXPRESSION AND THE WEB

Designers can be faced with a frequent blurring between the need for self-expression and communication, for monologue and dialogue, for ambiguity and clarity. Design usually serves a precise economical purpose based around the needs of a client – who typically initiates design work – and only purely experimental graphics can permit themselves the luxury of an internal rather than external raison d'être. For most graphic designers using print-based media, creating self-published material can be both time consuming and economically prohibitive.

This is, however, different in the case of web design, where these 'actes gratuits' (where the designer becomes their own client) have become a viable way for web designers to explore personal expressions of creativity, experiment with technology and raise their profile within the design community, due to the low costs of creating and distributing self-initiated sites. A large number of web designers therefore balance client-led commercial work with personal sites that allow them a creative outlet and can also be used as a promotional tool.

Joshua Davis is a New York designer who produces both public and private work on and off the web. Experimental sites are crucial to Davis's role as a web designer. These sites allow him to let off 'creative steam' and play with ideas that might not be appropriate in a client-led context, and they allow him a defined space for research and development for both design ideas and technical solutions that can be used in future projects for clients.

WWW.ONCE-UPON-A-FOREST.COM
Nº 10: Split/ Vis Croatia
Designer: Maruto (aka Joshua Davis)
2003

WWW.PRAYSTATION.COM
Rice paper, Kamakura
Designer: Joshua Davis
2001

Since 1997, Davis has produced two sites that have allowed him to explore his personal creativity in different ways. Like a scrapbook, his site **WWW.PRAYSTATION.COM** has existed as a research and development space, with clear objectives to apply new design and technology solutions to a range of creative concepts.

Davis's second personal site, **WWW.ONCE-UPON-A-FOREST.COM**, launched in 2001 with a very different purpose: 'I wanted to create a space for myself where I could ask questions, but not necessarily want to find answers. A place where I could let my own imagination reign supreme – and these projects, however mundane or inspirational, could be whatever I needed them to be.' The site existed as the private world of Maruto, a fictitious 'narrator', and Davis used it to inspire his imagination and the imaginations of users. Importantly, he wanted to reject the formal structures of web design, instead creating a space of confusion and exploration: 'No easy, short domain name. No easy-to-use navigation. No instructions. No FAQs. No ads. No links. No tech support. No help. No answers.' [8]

WWW.ABNORMALBEHAVIORCHILD.COM
Designer: Niko Stumpo
2002

In a similar way, Italian designer Niko Stumpo's personal site, **WWW.ABNORMALBEHAVIORCHILD.COM**, serves a dual function: the first section of the site works like a scrapbook, filled with ideas and sketches. The second section is an interactive moving image piece that exists as an exploratory space for the visitor, without clear narrative or navigational structures.

Creating personal sites allows designers to explore the structural nature of the Internet, unpicking elements of interactivity, communication and multimedia without the constraints of the externally imposed aims and objectives of a client. The **SODACONSTRUCTOR** site, created by UK interactive design group Soda, is an online construction kit for building animated models. Stemming from the idea of creative play, the user is able to set their own goals as they experiment and create with Java script, becoming part of a 'Sodaplay community'. Sodaconstructor exists both as a playful creative environment for the user and as a site of experimentation and research for the design group.

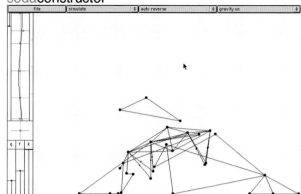

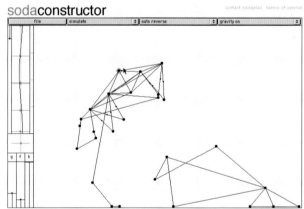

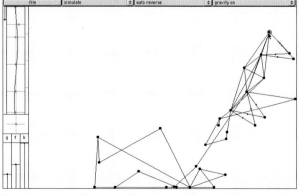

ARTISTIC VISIONS: THE BLURRING OF ART AND DESIGN BOUNDARIES ONLINE

The outcome of any design process, whether it is client-led or experimental and personal, should be considered in terms of the interplay between the designers' intentions and the users' needs and perceptions. At the same time these new practices highlight the fact that the role of the designer may embrace a range of identities – of narrator, mythmaker or artist.

The use of personal sites as spaces for creative expression undistracted by a direct relationship with client or consumer raises the question of whether boundaries are blurring between the role of artist or designer on the web. According to Philip Dodd, Director of the Institute of Contemporary Arts in London, this blurring of definitions is already taking place at a wider level: 'the notion that there is an essential difference (between art and design) is an historical fantasy…the traditional definition that art isn't functional and design is, doesn't work. Everybody is a designer and the word "design" ought to be rethought.' [9]

Are we seeing a convergence or a deregulation of creative boundaries in today's multilayered, multimedia world? In 2003 the National Design Triennial at the Cooper Hewitt National Design Museum in New York presented a simple mission: to show 'new ideas and future horizons across the field of current practice'.[10] Design staples including typefaces, transportation, lighting and logos were displayed alongside more idiosyncratic items: an artificial heart, a collection of smells and a theory of the universe. Onedotzero, a British digital moving image festival, purposefully explores the grey areas between art and commerce, client-driven and personal projects, featuring work by interactive design consultancies alongside digital filmmakers and architectural firms.

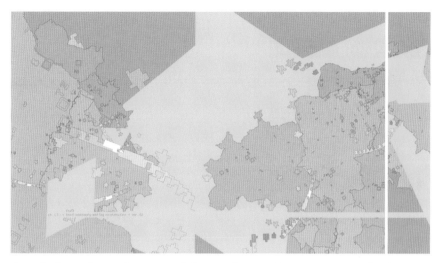

Are the boundaries between different forms of practice more fluid on the web? Graphic designers have always struggled with the fundamental tension between the role of 'artisan' and the role of 'artist'. For graphic design commentator and author Quentin Newark: 'Art is connotative, associative, implicative; it revels in ambiguities. Its function and form are inseparable. Design is precise, denotative, explicit. It is a mediation, a structure, a method.' [11] An artist is an individual whose work is concerned with self-discovery, which sets out to be unique, and which is formed through an experimental and speculative creative process. Most importantly, art has an intention and function that is internal to the work. The artisan or crafts-person, on the other hand, creates work that has an external or imposed function and intention.

Artists were among the first individuals to move into the newly created space of the World Wide Web from the early 1990s. When German artists Joachim Blank and Karl Heinz Jeron first became interested in the possibilities of the Internet in 1992, an effective graphical interface did not yet exist and the commercial potential for the medium was still to be discovered by many: 'In the early 90s, a lot of people said that the Internet wasn't interesting in terms of business. They didn't see the potential and the dynamics of the development going on. When we started out, all websites still had a grey background. That's why the medium wasn't at all interesting [at that time] to graphic artists and advertising agencies.' [12]

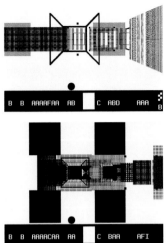

SOD
Artist: Jodi
2000

Net.art is a term that can be applied loosely to artistic projects that focus on the nature of networked media – predominately the Internet – in a formalist manner. Net.art explores key visual and formal concerns of the online environment. These include a self-reflexive interest in the structures and protocols of the Internet – from browsers and graphic interfaces to operating systems and programs – alongside a concern with the political, social and economic aspects of the medium – globalism, connectivity, immateriality, interactivity and equality.

SOD, a version of the early computer game **WOLFENSTEIN**, has been created by artists' group Jodi. Stripped of its surface design details, with the original elements – bricks, guards and animations – replaced by basic geometric patterns, this work unpicks the underlying codes and structures of the game. For Jodi creator Dirk Paesmans this formalist exploration is aimed at 'a better understanding of our own behaviour as users and players.' [13] Other work by a range of Internet artists explores the politicized nature of information distribution and the Internet: from the promotion of free information exchange through wide-ranging, high-performance free software – referred to as 'open source' – to wireless networking and global community action.

Italian artists 0100101110101101.ORG, for example, question the relationships between copyright, intellectual property and ownership on the Internet with a series of Situationist artistic activities. These have included the 'theft' of the online Net.art gallery **WWW.HELL.COM**, the posting of a copy of the Vatican website that included blasphemous statements and lyrics from boy band pop hits, and a work called **LIFE_SHARING**, which allows direct online access to the contents of the artists' computer.

Does the nature of the Internet itself create a unique visual language that has been adopted by both Net artists and designers? For a number of web designers, key Net.art concerns – an awareness of the underlying structures of the medium, the politicized nature of the Internet and a related refusal to be intimidated by technology – have informed their design work.

The creation of personal 'metasites' by designers (sites where the content is directed solely by the medium) in many cases mirrors the formal and visual qualities of Net.art: subverting users' expectations about the nature of interaction on the web, exploring the fear of technology failing, or investigating the social and political nature of networked technologies. American artist and programmer Brett Stalbaum has defined the typical elements of 'web formalism' in Net.art:

> Features… include user interface tricks (fake buttons and menus as examples), use and repurpose of data from outside the site elsewhere on the web…the alternate, extended, 'wrong' or 'creative' use of HTML or browser features to achieve unexplored effects... Stated far too simply but conveniently: web formalism is the use of the web for the web's sake…[14]

In many ways the same stylistic tropes can been seen in design 'metasites', including **WWW.PRAYSTATION.COM** by Joshua Davis, James Paterson's **WWW.PRESSTUBE.COM**, **WWW.DEXTRO.ORG**, or **WWW.SOULBATH.COM** by London-based design group Hi-ReS!.

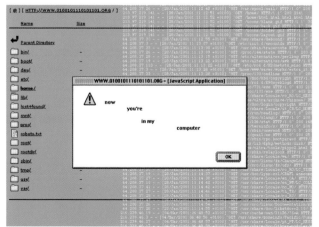

LIFE_SHARING
Artist:
0100101110101101.ORG
2001

WWW.DEXTRO.ORG
Image: m104
Designer: Dextro
Format: Screenshot from interactive
shockwave animation
2002–3

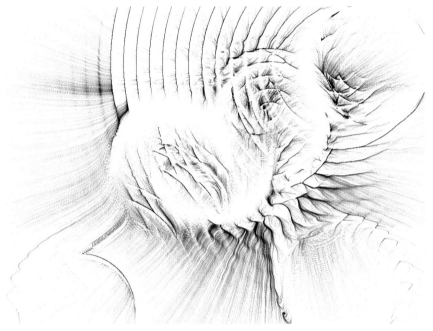

Hi-ReS! was founded in 1999 by Florian Schmitt and Alexandra Jugovic, and they have since created a number of award-winning commercial sites, including websites for the films 'Requiem for a Dream' and 'Donnie Darko' and **WWW.THETHIRDPLACE.COM**, a creative online space for Sony's Playstation 2.

Hi-ReS! aim to keep the user experience at the heart of their research and development process, subverting users' expectations with a tendency towards chaos and malfunction, as Florian Schmitt explains:

> I always have the user in mind. When we created **SOULBATH.COM** I thought about how scared people using the site would be – technology going wrong can be really frightening. We want to challenge people's preconceptions. We definitely don't try to be difficult for the sake of it – we hope that people will find the process of looking for answers rewarding. I think it is good to feel that there is stuff left to discover.

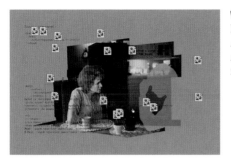

WWW.REQUIEMFORADREAM.COM
Design Agency: Hi-ReS!
Artistic Direction, Design and Production:
Alexandra Jugovic, Florian Schmitt
2000

In their website for the Darren Aronofsky film 'Requiem for a Dream' Hi-ReS! use a similar visual language to Net.art to undermine the security of the user interface and to express the unpredictable consequences of technology failing. The site slowly decays – with increasingly corrupted images, disrupted text and unexpected links – to mirror the gradual unravelling of the lives of the characters within the film, creating an awareness of just how fragile notions of 'choice' and 'control' really are.

However, although in many ways the website for 'Requiem for a Dream' and other sites created by Hi-ReS! share a common visual and formal language with Net.art, Florian Schmitt does not consider their work to be art: 'I couldn't care less, but I wouldn't want to burden ourselves with calling ourselves artists.' While the visual language may be the same, the sites that Hi-ReS! create serve a specific and defined purpose: as promotional websites for films or music bands, or as engaging and interactive online spaces for companies to enhance their brand identity – from Lexus and Mitsubishi to Sony's Playstation 2. Unlike Net.art, personal and client-led sites created by web designers cannot fully escape the concerns of the commercial world. The work exists in a dialogue – creative expression feeds into future client-led projects, old work informs new and personal visions become allied to new brand experiences.

COLLECTIVE VISIONS:
COMMUNITIES AND COLLABORATIONS ONLINE

While some designers have adopted the visual and structural concerns of Net.art, others have been inspired by the growing awareness of the political, social and economic implications of the Net – a vision of the Internet as a social space, providing opportunities for social and political critique and the formation of a collective, community-based approach. These roles – of activist, collaborator, and community facilitator – may alter the future of interactive design just as significantly as continuing technological innovation.

New media critical debate has traditionally focused around two distinct and opposing paradigms of virtual space. The first is the notion discussed earlier (see page 16) that the body and identity of the viewer disappear as the relationship between vision and the screen become all-important: 'nothing could be more disembodied or insensate than…cyberspace. It is like having your everything amputated'. [15] For theorist Kevin Robins, the pre-eminence of vision and the concurrent rise of new technologies allow the individual to avoid contact with external reality; instead, 'technologies function to mediate, to defer, even to substitute for interaction with the world'. [16] This representation of a neutral and depersonalized space denies the possibility for any embodied experience or social connectivity.

WWW.ANTIWARGAME.ORG
Design Agency:
Futurefarmers
2003

KOSOVO ELF
Design Agency:
Futurefarmers
1999

The contrasting view to this is that as a constructed media (created by people living in the 'real world', for people who live in the 'real world') it continues to embody the social, cultural and political practices that surround it. The space beyond the screen is not neutral – a utopian site disconnected from the 'real world'. Instead it can represent, embody and critique existing cultural structures or create new ones. In addition, this view posits the idea that the interactive nature of the Internet allows the creation of new kinds of connections and relationships between individuals and communities. This notion of a 'liminal' rather than utopian space – a space between the existing ordered social structures and relationships in the 'real' world – has been caused in part by the nature of the Internet itself, a decentred global matrix embedded with a culture of downloads, peer-to-peer networks and open source and freeware Net communities.

Within this framework, designers hold within themselves the power both to reflect the society around them and, if they choose, to actively shape and refine these concerns. As French designer and curator Philippe Apeloig states: 'Design is fundamentally idea-oriented, and designers carry profound influence in their power to shape and communicate cultural concepts.' [17]

Since 1995, San Francisco-based design group Futurefarmers have created personal work that explores these social, political and cultural concerns. Amy Franceschini and colleagues have built a creative fusion of playful Japanese cuteness, wryly hip San Franciscan irony, and a unique sensibility for virtual materials and information architecture, blending a successful commercial practice with an interest in notions of community, sustainable environments and collaborative working practices – an idea that they call 'communiculture'. Futurefarmers' self-initiated projects critique elements of contemporary society, and include interactive games such as **ANTI-WAR GAME** and **KOSOVO ELF** that, in a non-threatening cartoon style, respond to recent military activity by the US.

WWW.THEYRULE.NET
Design Agency:
Futurefarmers
2002

WWW.1111111111111111111111
11111111111111111111111111111
111111111.COM
Designer: Kenneth Tin-Kin Hung
2002–

THEYRULE is an online project that was featured in the 2002 Biennial Exhibition at the Whitney Museum of American Art. Created by Futurefarmers designer Josh On, TheyRule investigates corporate power relationships in the US, allowing users to browse through a variety of maps that function as directories to companies such as Pepsi, Coca-Cola and Microsoft. TheyRule depicts the connections between companies through diagrams of their power structures. Users can run web searches on CEOs, the donations they have made and their companies. They can also add to a list of hyperlinks relevant to that company or person or save a map of connections, complete with annotations, for others to view.

Based on C. Wright Mills's book 'The Power Elite' (1956), which documented the interconnections between the most powerful people in the US, TheyRule employs the features of networked technologies, such as dynamic mapping, hyperlinking and instant searches, to create its own networks of power systems.

'FUGITIVES' EXHIBITION
The Riviera Gallery, Brooklyn, New York
2003

TheyRule refers back to the early promise of the Internet as a 'democratizing medium' rather than a marketing tool, and it is this aspiration that has inspired other designers to use the web as a tool for social and political commentary. US-based designer Kenneth Hung, for example, uses his personal site, **60X1.COM**, as a site for free expression, critiquing the military activities of George W. Bush's government through a series of flamboyant and satirical set-pieces.

These notions of the web as a democratizing medium and a space of free expression have helped to engender a sense of community between designers working online. Design portals, created on a non-commercial basis, have become a key way of supporting this sense of a community of designers: from well-known sites such as **KALIBER10000** (**WWW.K10K.NET**) and **WWW.THREEOH.COM**, to London-based **WWW.LINKDUP.COM** and Australian site **WWW.DESIGNISKINKY.NET**.

A collaborative approach is something that defines many online projects. Between 1997 and 2001 the **REMEDI PROJECT** – an online interactive art gallery created by US designer Josh Ulm – presented twelve exhibitions of experimental work from over sixty digital practitioners, including many of the leading names in the web design industry. Other self-initiated projects have included Matt Owen's **CODEX SERIES** – a series of CD-ROMs that Owens has termed 'somewhere between a compilation and a digital fanzine' – providing a space for creative practitioners to showcase their work in a creative and supportive environment, while a growing number of online magazines are showcasing the work of artists and designers.[18] These collaborative ventures are not limited to online projects. Both Futurefarmers and Matt Owens have engaged in offline ventures. For Matt Owens this move 'offline' has become more permanent with the launch in 2003 of **RIVIERA**, a gallery space co-founded by Owens in Brooklyn, New York.

WARHOLISER
Design Agency: Poke
Client: Tate Modern
2002

The shake-up of the web design industry that followed the dot-com slump in 2001 has played a role in reinvigorating this collective approach for designers in the workplace. The economic downturn at the turn of the Millennium preceded the closure of a number of high-profile large interactive design consultancies, including well-known companies such as Deepend UK, Razorfish and Kioken in the US. The regrouping of the new media design industry has, along with a number of other factors, including the growing convergence of new media, telecommunications and broadcast technologies, and the increasingly multidisciplinary nature of graphic design and moving image work, affected the way designers are seeing themselves and the choices they have in terms of workplace opportunities.

In the early heady days of the Internet, design collectives such as Antirom, and subsequently Tomato Interactive, were an influential part of the new web design industry, forerunners to the explosion of large-scale digital design consultancies and agencies. Increasingly, freelance web designers are once again protecting themselves against a fluctuating economic environment by joining together to form collectives such as The Embassy of Code and Form, a worldwide network of designers, developers, writers and others founded in 2001. Co-founder Patrick Sundqvist explains: 'things started to crumble around us as economies went into recession and many of us felt we were not doing the work we wanted, either when it came to the type of work we did, or to the quality of the final output. The whole idea behind this network was to create a working environment where we could share work, help each other out on projects, [and] form working groups of the best talent needed for the task at hand.' [19]

WWW.13AMP.TV
Design Agency: Friendchip
Designer: Anthony Burrill
Programmer: Kip Parker
Client: 13 Amp Recordings
2001–2

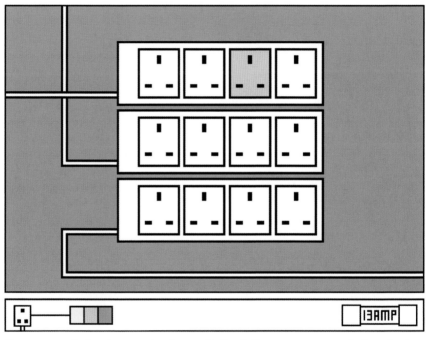

Because of the increasingly multidisciplinary nature of the design environment, a collaborative approach can be a vital part of the success of a design agency working today. London-based design consultancy Poke (founded in part by ex-Deepend staff following the closure of the London office of the company) are able to take a multidisciplinary stance by nurturing a series of internal and external collaborations: internally they assemble appropriate one-off teams, using freelancers as appropriate, while externally they also benefit from close relationships with advertising and media agencies Mother and Naked. The designers at Poke do not wish to specialize in any one particular aspect of interactive media: 'We do a broad range of work through advertising, marketing communications and the "TV space"…we think of it as holistic thought and specialist delivery.' [20]

'PAPER CLIP' AND 'CHIP FORK'
Objects of Limited Value
Designer: Anthony Burrill
2003

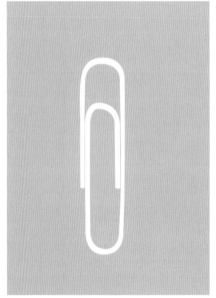 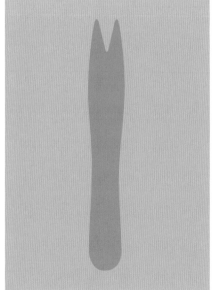

For London-based designer Anthony Burrill, the most effective way he has found to maintain a multidisciplinary approach and ensure a balance between client-led and personal projects has been to create a series of 'personalities' – through a number of long-term collaborations. As 'Anthony Burrill' he blends personal print-based projects with commercial graphic design work for clients such as London Underground and advertising agency Kesselskramer (including poster campaigns for Diesel and the Hans Brinker budget hotel in Amsterdam). He also works within two different collaborations: creating interactive projects with programmer Kip Parker as Friendchip and, as Multiplex, designing moving image projects in collaboration with Paul Plowman and Malcolm Goldie.

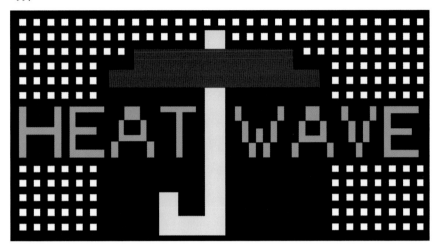

Burrill graduated from the Royal College of Art in London in 1991 having focused on book design and simple moving image projects. Burrill met Kip Parker when he was seventeen. The first online project they created together was a small website called **GETJETSET** in 1997. Since then they have worked together on a number of commercial web projects, mostly related to the music industry, including websites for Kraftwerk, the 13 Amp record label and French band Air.

For Burrill, the act of collaboration is extremely important. While he and Kip Parker retain specific roles within the projects they work on together – Burrill as art director, Parker as programmer – the creative process itself is built upon 'bouncing ideas' back and forth. By maintaining a regular level of commercial graphic design work – which Burrill calls his 'bread and butter' – he is able to view both Friendchip and Multiplex as more personal projects, giving him a freedom to concentrate on self-initiated work and to select a small number of client-led projects that interest him and allow him to maintain a level of creative vision and control.

At the heart of the interactive design process lie these dialogues: between personal and client-led visions, between the design process and wider social contexts and, most importantly, within the interactions between the user and the content beyond the screen.

40–65

STYLING THE WEB

GRAPHIC DESIGNERS ARE LOOKING TO REVEAL THE HUMAN ASPECTS OF DESIGN – THE SMUDGE, THE MARK, THE TRACE OF A HAND OR GESTURE, THE BEAUTY IN IMPERFECTION.

BOSSANOVA LP SLEEVE
Pixies
Art direction and design:
Vaughan Oliver
Photography: Simon Larbalestier
Model-making: Pirate
1990

Among the myriad of stylistic choices made by web designers in recent years, two distinct visual languages have emerged. The first embraces the technological narrative of the computer, forging a new aesthetic from icons, pixels, codes and grids. The designers who use this language utilize stylistic devices from the early years of gaming and computers, evoking a wry nostalgia for a techno age barely reaching maturity. The second tendency has been to build bridges between real and virtual worlds, employing a 'handmade' aesthetic and appropriating vernacular and historical references, blending the new with the old, the analogue with the digital.

Such stylistic exploration goes beyond the search for novelty. In the disembodied space beyond the screen, designers are searching for ways to recognize the role of personal experience and engagement in the design process; a way to weave traces between the digital realm of creative possibilities and the tactile, tangible elements of the real world. There is a growing desire to bear witness to the physical act of making. Graphic designers are looking to reveal the human aspects of design – the smudge, the mark, the trace of a hand or gesture, the beauty in imperfection.

The roots of current web design practice can be traced back to the stylistic experimentation of the 1980s, when graphic designers such as April Neiman, Dan Friedman, Neville Brody and Vaughan Oliver used visual strategies to investigate and undermine the conventions of Modernist design. These postmodern graphics employed the decentring, layering and manipulation of images and typography, and the use of design tropes that had previously been rejected or disregarded in the Modernist movement. These included the use of non-functional decorative elements, the appropriation of vernacular or street styles and the use of historical references from different periods, styles and cultures.

WWW.ENTROPY8.COM
Designer: Auriea Harvey
http://entropy8.com/greatest_hits
1995–9

These trends converged with the introduction of computer-assisted design, offering designers a new vehicle for visual and technical experimentation. The unfolding strands of postmodern design became intertwined with electronic capabilities. For the first waves of web designers, the visual language of postmodernism offered a parallel to the nature of the Internet itself. The rejection of conventional notions of visual syntax, hierarchies and imagery fitted well with the fluid and non-hierarchical nature of virtual space and the multi-linearity of the links, nodes and networks of hypertext. At the same time, first CAD, and later the World Wide Web, provided the tools and platform for pioneering work.

While designers like John Maeda turned to examine in greater detail the structural nature of the Internet, for others the web offered a chance to play with the visual conventions of graphic design. Websites that exemplified this tendency included Auriea Harvey's **WWW.ENTROPY8.COM**. From 1995 to 1999, this site provided a space for experimentation with an unorthodox use of HTML, animation and visual narratives using gothic styling and multiple layering techniques.

Other web designers were inspired by the loose, anti-grid, 'anti-style' of designers like David Carson and his underground magazine 'Raygun'. Designers in the late 1990s like Miika Saksi transferred elements of this 'lo-fi' design style onto the web by featuring thin strips of information, intentional scrolling, smudges, misaligned fonts and frozen screen layouts. These elements attempted to create a more 'analogue' feel to web design.

WWW.DESIGNGRAPHIK.COM
Designer: Mike Young

GUNS, GIRLS, CARS, ROBOTS AND MONSTERS

Technological innovation has driven stylistic choices from the birth of web design. The technologically induced optimism of the early to mid-1990s, and the new possibilities offered by design tools such as 3D modelling software, has seen a wave of futuristic online landscapes spreading across the web. Designers like Mike Young embraced a shiny, technologically progressive digital vision of sharding 3D vectored landscapes with cryptic navigational elements and tiny text detailing. These 'futuristic' designs became, for many, the pervasive face of web design at the beginning of the twenty-first century.

PAC-MAN IMAGE
Designer: Mr Toru Iwatani
1980
© 1980 Namco Ltd.

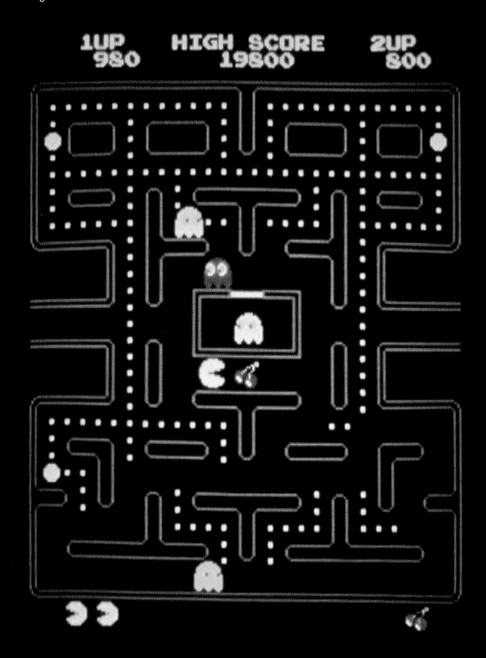

WRITER'S BLOCK
Designer: Jon Bell
lot23.com
2000

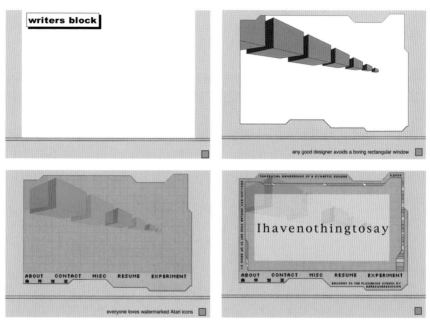

Designer John Hoyt outlines some of the stylistic tropes that have originated from this design style and been adopted by many web designers, both successfully and unsuccessfully: 'Forty-five degree angles, pink and grey colour schemes, and the ever elusive small text. Pop-up windows, scrollable areas, css [cascading style sheets], layers and action script…Flash-based projects, dhtml widgets, and funky audio plague the sites I look at daily.' [1]

While the increasingly standardized nature of current web design has been satirized on websites including **WWW.LOT23.COM**'s 'Writer's Block', other designers have rejected this relentless pursuit of new technological visions, and are instead looking back – creating stylish nostalgic references to early gaming and pop culture. From **PONG** to **PAC-MAN**, designers have revisited the flat 2D graphics and blocky pixels of the early computer gaming era of the 1970s and 1980s. Rejecting the current gaming industry trend towards a 3D hyper-realistic graphic style, designers are instead attracted to earlier 2D forms inspired by classic arcade games like **SPACE INVADERS**; pixelated environments reminiscent of **SIMCITY** or **ZELDA**; or the reinvention of 'cute' animated characters from early platform games.

WWW.INTAIRNET.ORG
Design Agency: Friendchip
Designer: Anthony Burrill
Programmer: Kip Parker
Client: Revolvair
2004

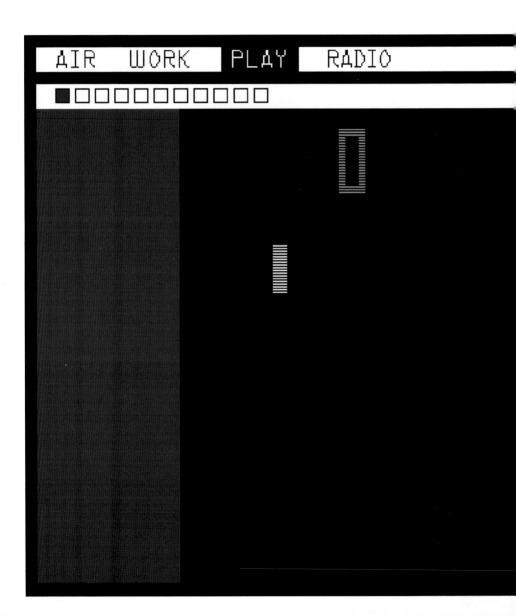

When Friendchip were commissioned by French band Air to create the website **WWW.INTAIRNET.ORG**, designer Anthony Burrill and programmer Kip Parker used the opportunity to create a 'play section' that could explore the band's aesthetic while recreating the feel of early computer graphics using a minimalist and humorous approach. Games include, in Burrill's words, 'a bad version of Pong' inspired by the video to Air track 'Kelly Watch the Stars', as well as an interactive cover version of the song 'Sexy Boy'.

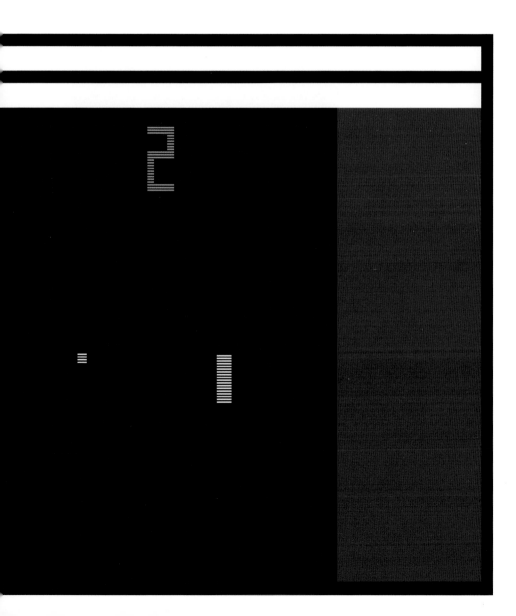

WWW.STRIDE-ON.COM
Design Agency: Fibre
Client: Overland group
2001

TXT MTCH
Design Agency: Fibre
Client: Motorola/The Fish Can Sing
2002

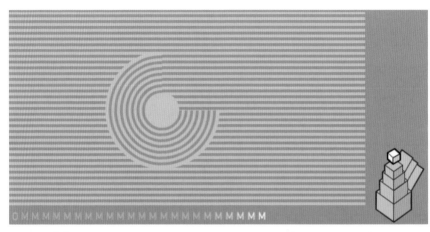

Design group Fibre have created a series of retro-inspired games for client-led projects. These include a 'chill-out zone' for users of the Stride footwear website **WWW.STRIDE-ON.COM**, which features Pachinko games and Japanese-inspired sand raking alongside other relaxing pastimes. For Fibre designers Nathan Lauder and David Rainbird, games can provide 'sticking power' for users on a website: 'Games are fun. They involve people. A key desire for companies is to encourage people to return again and again to their website. You need something that people will go back to for pleasure, and games can provide that.'

Fibre's interest in gaming is not limited to arcade games. In 2001, Fibre were commissioned to create a marketing campaign for the digital festival What Do You Want To Do With It? at the ICA in London. Fibre included a 'Draw' section on the festival website, which acted like a 'souped-up, on-line Etch-a-Sketch', letting users create their own pictures and post them on the site. At the same time, Motorola invited Fibre to design a game for the festival and they produced **TXT MTCH**, a game that blended mobile phone texting with Space Invaders. Played online or at a kiosk, the game required players to translate a series of words from Standard English into increasingly complicated 'text speak'. Txt Mtch proved popular with both the public and critics. Nominated for a British Design and Art Direction (D&AD) Award in 2002, one contact at Motorola aptly summed up the finished product as 'FKNG GR8'.

German design company eBoy distil popular culture into a pixellated environment of guns, girls, cars, robots and monsters. Describing themselves as the clones of 'email, Apple and the Beastie Boys' mixed with an added '10,000 volts',[2] eBoy produce both online and print-based graphic work rich in colour and inventive in design.

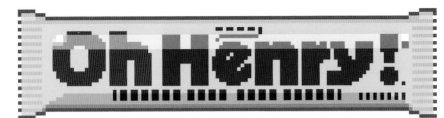

PLICAS PROJECT
Design Agency: eBoy
Designed for 'Super, Welcome to Graphic Wonderland',
Thomas Brussiger and Michael Fries
2002

BOY MEETS PIXEL
Designer: Craig Robinson
2002–3

According to eBoy designers Steffen Sauerteig and Kai Vermehr, their love of pixels 'comes from working for the screen. We [make] icons, websites and software surfaces. To get really sharp and clean results you have to put in every pixel by hand.' [3] This method provides them with a stylized way of interpreting their surroundings:

> You're permanently forced to abstract because you have a limited number of units [or] pixels for the object you are trying to visualize. You can't just draw freely as you might do on paper. If you want the lines to look clean, you're restricted to certain angles...this way every single thing you create... is an interpretation – not a copy. [4]

ECITY
Design Agency: eBoy
Image created for 'The Face', May 2002

In the same way that eBoy are able to use pixels to distil culture and society to its basic elements, for example through their recent **PLICAS** project, British designer Craig Robinson's pixel portraits in his **MINIPOPS** and **BOY MEETS PIXEL** series reduce personalities to one or two visual tropes.

EBoy first became known for creating a series of online eCityscapes – virtual metropolises that grow organically and in staggering detail, reminiscent of the graphic world of computer game SimCity. However, computer and arcade game imagery from the 1980s are only part of a range of pop culture influences on eBoy's design style. Most importantly, eBoy refuse to take design too seriously: 'Grow up? In what sense? Should we do "serious" stuff? …You don't just stop loving toys, comics, Disney…There is no growing up anymore…that was twenty years ago. Nobody does it anymore.' 5

THE OLD AND THE NEW:
HUMAN IMPRINTS ON A DIGITAL WORLD

'The new generation of young designers is realizing that everything comes out of the computer looking very clean. Even if it's raw, it's clean, so they are beginning to move towards other processes.' [6] Huw Morgan, Graphic Thought Facility

A new generation of designers is choosing to reject the homogeneity of the polished, 'finished' surface available through digital technology and is instead choosing to experiment with hand-crafted, analogue production techniques and a range of organic, decorative and illustrative styles. These designers, inspired by wider movements in graphic design, are looking to engage in a more direct and personal way with their work. Value is placed on generating connections on a 'real' rather than 'hyper-real' scale – with human perspectives and human imperfections. As author Andrew Blauvelt writes: 'This is an inverted world where the ordinary stands out from the crowd as a distinctive gesture…what seems trivial and tangential becomes essential, like so many bits and pieces of data in the detritus of the information age.' [7]

Interest in handmade and hand-drawn processes in particular stems from a desire to find new ways of working in reaction to dominant digital art forms. Publications like Anne Odling-Smee's 'The New Handmade Graphics' have shown that, for some, these analogue practices seem to offer an alternative vision to that offered through the globalized mass-production techniques linked to the digital desktop revolution. For Anne Odling-Smee, the raised level of interest in traditional design methods is 'a reaction to technology since the digital revolution of the 1980s…and secondarily, to things like globalization and environmental concerns which are themselves consequences of technological developments'. [8]

OXFORD RAGWORT
From the series 'Nourishment'
Artist: Michael Landy
Courtesy Maureen Paley/Interim Art
2002

This growing interest in traditional techniques reflects the cyclical nature of fashion trends. For designer and lecturer Stuart Bailey, an illustrative element existed in graphic design during the 1980s and 1990s – from the club flyers of Jason Brooks and Graham Rounthwaite to the slick, aspirational illustrative style of 'Wallpaper' magazine. However, the early years of the twenty-first century have seen an increasingly receptive audience (both critically and commercially) attracted to this style's apparent newness, '…but only because it is different from what went before – which may be digital-looking, which may have replaced punk-looking, which may have replaced neutral-looking…'. For Bailey, '…the only work that really interests me is that which starts with the content, not the style'. [9]

Recent critical debate has placed this 'revival' of handmade processes in graphic design within a wider cultural context, allied to related trends in a number of creative fields – including a resurgence of interest in craft techniques and customization in fashion, lo-fi music production and fine art. A number of artists, including Michael Landy and David Hockney, have spoken of a rediscovery of drawing, painting and etching techniques in their recent work.

Within graphic design, this shift of focus has seen a revival of decoration and ornamentation – from stylized rococo embellishments to more organic, natural forms. Although much of this design style is inspired by pre-digital techniques, for Alice Rawsthorn, Director of the Design Museum in London, the most interesting work stems from designers who are able to combine ornamentation with an awareness of the possibilities that technology can offer: 'the best examples are produced by designers who…harness advanced technologies to mass-manufacture work with a level of richness and intricacy that could hitherto only be achieved by hand-craftsmanship'. [10]

POSTER FOR BLOW UP MAGAZINE
Designer: Nando Costa
www.hungryfordesign.com
2002

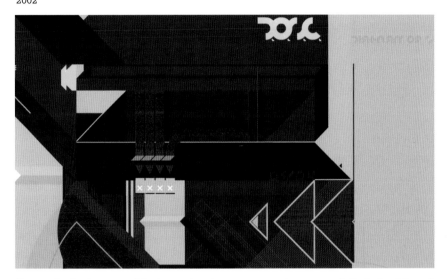

For web designers, this interplay between analogue and digital, old and new, presents exciting possibilities for creating new personal visions online. Brazilian designer Nando Costa actively exploits the crossover between different media, blending a riot of colour and organic and natural forms with more mechanical imagery. Ben Radatz of US motion graphics studio mk12 says that Costa's talents lie in using a versatile range of skills but without drawing attention to the medium itself: 'His work in print could be an animated short film or a Flash site, and vice versa.' [11] Costa says his Brazilian background has played a key part in the evolution of his style: 'What's interesting is that Brazilian designers have a large interest in the arts in general and most of them have previously been artists…In the US, most designers have been introduced to design after only a very brief experience with drawing, for example. The experience I've had has definitely influenced how I design.' [12]

For designers like Nando Costa, taking a cross-disciplinary approach and blending a range of analogue and digital design techniques can provide new channels for exploration – creating dialogues that can involve both online and offline practice.

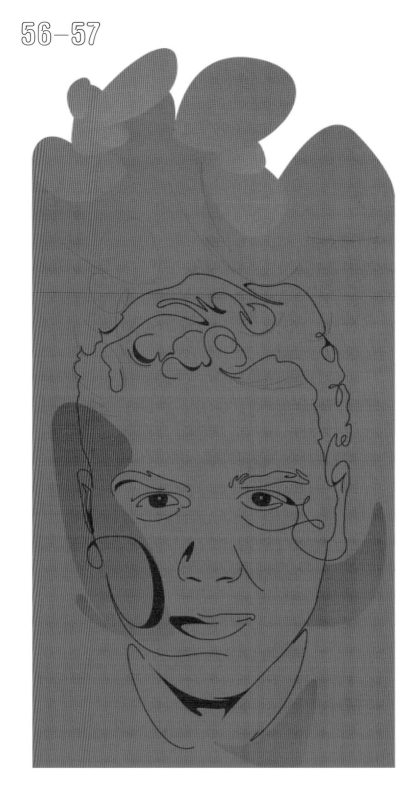

←—

**SPLASH PAGE FOR
WWW.DESIGNISKINKY.NET**
Designer: Nando Costa
www.hungryfordesign.com
2002

←

WWW.KONEISTO.COM
Design Agency: Syrup Helsinki
Antti Hinkula and Teemu Suviala
Client: Koneisto Production
2002

Designers Antti Hinkula and Teemu Suviala from Finnish design agency Syrup Helsinki use a wide range of design tools – both digital and non-digital – to create print-based and online work: 'Freehand, PhotoShop, Aftereffects, hands, Flash, camera, Streamline, Final Cut Pro, paper, pencils, watercolours, scissors, ink, spray, brush, food, scanner, fax, printer and other machines and anything else.' They describe their creative process as 'visually composing': sampling and mixing different visual elements together. When designing the brand identity and marketing campaign for the annual Koneisto Festival of Electronic Music and Arts in 2002 and 2003, Syrup used the idea of 'an album of rhythms and beats and melody built and mixed of graphic elements, shapes and colours'. Although visually quite different, the design identities for the Koneisto Festival in 2002 and 2003 reflect a willingness to experiment with a range of design tools and techniques, overlaying typography, photography, animation and illustration to create a melting pot of colour and pattern.

The 2002 Koneisto Festival website blends repeating decorative elements with slogans and cut-and-paste text soundbites. The stylized monochrome floral illustrations form intricate patterns through the site – an Art Nouveau styling that springs to life through animation and sound, creating different 'rhythms, beats, tempos and loops'.

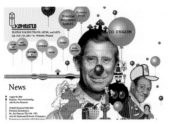

WWW.KONEISTO.COM
2003

For the 2003 Koneisto Festival website, Syrup used these methods of overlaying text and images, and contrasting illustration with photography and animation, to create a vibrant explosion of colour and image:

When we started working on Koneisto 2003, it was the time when the USA attacked Iraq…We came up with the slogan 'Come Together', which has many different meanings…That slogan became the main idea behind all the graphics; we wanted to create a family of different characters which would perform together in our designs. We combined hand-drawn pencil portraits of famous, but in some way weird characters, like Prince Charles, Bill Cosby, Monchichi dolls or Einstein. Then we illustrated a family of vector style characters…like Popeye, Minnie Mouse, an Apple logo with three eyes…My Little Pony with stripes, a guy made of Michelin tyres…We used a rich colour palette and wanted the visuals to look very joyful. We used colourful 3D balloons to make it even happier.

At the heart of these new 'handmade' expressions is the subtle, expressive nature of illustration: 'Illustrators have their eyes fixed on the jugular of the Zeitgeist' says Gerard Saint, director of creative agency Big Active. [13] The act of drawing provides for some the clearest and simplest approximation to the physical act of creation. For author and educator James McMullan, drawing is 'informed reflex': '…a physical act which puts us in touch with how we really experience space and form …Without drawing there would be no way to meld the world of the rational and the world of the intuitive.' [14]

The blend of animation and interaction makes the Internet a natural home for experimentation with illustration. James Paterson is a Canadian illustrator and designer who creates a kinetic riot of flowing, playful and roughly sketched line drawings through his personal website **WWW.PRESSTUBE.COM** and **WWW.INSERTSILENCE.COM**, an ongoing collaboration with musician and programmer Amit Pitaru.

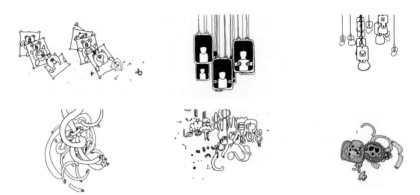

Paterson uses the Internet 'like a sketchbook', transferring the skills and sensibilities of figure drawing into movement and code. Working with computers has altered the way in which Paterson draws, both online and offline:

> Recently I realized that the pen I use in my sketchbook (a pilot tech pen) is as close as I could possibly get to a single pixel scribbler in Flash or PhotoShop. Without recognizing why, I just sort of got hooked on a pen that is compatible with the work I do on the computer. So when I am drawing all over the place, in a bar or on a bus or at home, all of the work I am doing in my book is totally interchangeable with my computer work.

In the same way, Paterson finds that 'the organization and style of writing code has overflowed into my drawing process. It has changed the way in which I approach drawing and what interests me about it. It is almost as if I run programmes sometimes when I draw now.' [15]

Both Paterson and Pitaru are interested in embracing the often sterile process of coding in an emotionally expressive way. Paterson explains: 'A lot of work for me is trying to communicate the way that I perceive music… [in the same way] my drawings are my way of filtering everything through myself. Anything I see or experience that makes an impact on me, I try and feed to my drawings.' [16]

CREATIVE VISIONS: STYLISTIC EXPLORATIONS AND PERSONAL VOICES

The opportunities and benefits of taking a cross-disciplinary approach to online creativity are perhaps best expressed through the growing online magazine scene. The magazine sites bear witness to the potential for mixing media – photography, illustration, text and multimedia – and the positive value of creating dialogues between digital and analogue processes.

Online magazines have become a global phenomenon, featuring the work of designers and artists from Iceland to the Philippines and from Brazil to Japan. Like photocopied fanzines before them, these independent collaborations are cheap to produce and offer a freedom of expression and creative experimentation. Since May 2002, Andy Simionato and Karen Ann Donnachie have produced **THIS IS A MAGAZINE**, one of the best-known online magazines. For Simionato, the collaborative nature of a magazine format allows a freedom to explore multimedia capabilities alongside the traditional models of printed communication: 'We are able to explore the fringes of visual communication because of the hybrid nature of the Internet – sound, visuals and movement…whilst maintaining the traditional rhythms and frameworks of the magazine format.' [17]

Although for the most part existing solely in a digital form, these magazines take their design cues from print-based publishing. Content is laid out as 'spreads', you 'turn the page' to navigate, and the magazines often include sound effects of pages turning or feature virtual borders, margins and paper folds. Brian Taylor, editor and publisher of **DODGE** online magazine, feels these design restrictions can act as a structuring device: 'the strength of Dodge is its underlying rhythm. Set against the meter of design guidelines, individual improvisations, while amazingly varied…seem part of a greater whole.' [18] The blend of new and old technologies in a multimedia context can add a richness of content to these sites. Within **F MAGAZINE** for example, innovative content sits as part of a rich media experience that includes interactive animated advertising and music video streaming.

WWW.THISISAMAGAZINE.COM
ISSUE II, 'What is art?'
Designers/Editors: Andy Simionato and
Karen Ann Donachie
Illustrations, below, by Andy Simionato, of
the Neen dogma written by Miltos Manetas
2003

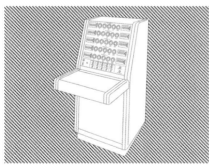

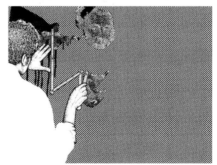

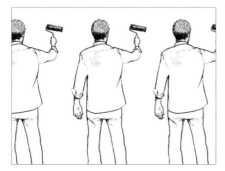

POSTER FOR PRINT 'EM
Designer: Takeshi Hamada
Photo: Per Zennström
Client: Mitsubishi Paper Mill Ltd
2003

WWW.TIGERMAGAZINE.ORG
Issue 26
Designer: Takeshi Hamada
2003

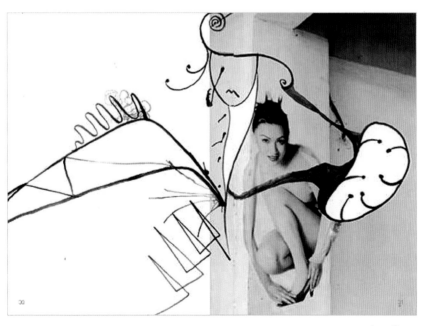

While studying in Germany in 2000, Japanese design student Takeshi Hamada created an online project in Flash called **DIE DAUMENKINO** (meaning 'thumb cinema'). Using this simple format, Hamada produced his first self-published work online, retitled **TIGER** after his nickname.

In his print-based work Hamada creates an original vision of graphic design that marries strong photographic elements with delicate tracings of line drawings and watercolour, creating decorative patterns of line, text and subtle colour. Hamada's initial interest lay in illustration: 'when I started working as a graphic designer I didn't know the difference between graphic design and illustration. Maybe it was because the art school I graduated from is in the Japanese countryside! As I originally wanted to be a comic artist or illustrator, I decided to be a graphic designer since I thought the graphic designer's main job was to draw.'

A key aspect of the creation of Tiger lies in the collaborative approach Hamada takes to its creation – commissioning work and accepting contributions from viewers while at the same time maintaining a coherent vision that features a selection of his own graphic work. Tiger allows Takeshi Hamada a freedom of expression and a freedom to slip between the boundaries of different media: he calls Tiger a 'play magazine' – on the borderline between a publication, an art space and a site for personal enjoyment and experimentation.

For other creative practitioners, the Internet provides an opportunity to create non-linear personal narratives that blend illustration, text and moving images. One of the signs of a young medium is the proliferation of work exploring the technical and visual parameters of an art or design form, without necessarily a concomitant richness and diversity of content. As designing for the web matures, practitioners are beginning to explore the narrative possibilities of the medium, and their additional roles as 'storyteller' or 'mythmaker'.

WWW.NOBODYHERE.COM
Artist: Jogchem
Niemandsverdriet
1998–

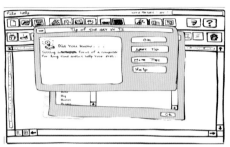

WORD PERHECT
Artist: Tomoko Takahashi
Commissioned and produced by e-2.org
2000

Tomoko Takahashi's **WORD PERHECT** asserts an idiosyncratic and deeply personal human voice by creating a hand-drawn version of the ubiquitous word processing program. As word processing software becomes ever more advanced, correcting syntax and spelling errors, these familiar programs begin to impose a standardized corporate language onto our writing. Takahashi has produced her own fully functioning online version that undermines this dehumanizing process while reclaiming rights as an individual over the impersonal nature of the machine.

The Internet enables new kinds of interactive narrative forms through its matrix-like structures of interconnected links and the concurrent blurring of boundaries between virtual 'author' and 'reader'. Collaborative sites such as **BORN MAGAZINE**, an experimental forum that marries literary arts and interactive media, encourage non-traditional approaches to storytelling, and forge new paths across and between different genres of creative practice.

NOBODYHERE, created by Jogchem Niemandsverdriet and supported by the Netherlands Foundation for Visual Arts, Design and Architecture, is a site that successfully uses the underlying nature of the Internet – elements of non-linearity, interactivity and multimedia – to create an ongoing journal of connected thoughts, works in progress and personal commentary that blends sketches, photography and interactive animations. The viewer is drawn on a seemingly random journey through Niemandsverdriet's good intentions, failures and desires: the detritus of the everyday.

CAFFEINE SOCIETY
Devised and directed by Mario Cavalli
Client: iDN Magazine
2001

In a commercial context, designers are using these techniques to engage the viewer with narrative elements and character development, allowing the viewer the opportunity to construct stories and personalities online. Within a predefined virtual environment, websites like the **LEXUS MINORITY REPORT** or Diesel's **PROTOKID** allow the user to take on a role within an interactive narrative, and to make a number of choices that alter the progress of the story. **CAFFEINE SOCIETY**, Mario Cavalli's award-winning splash page for iDN magazine in 2001, is an early example of a site that adopted a narrative structure to create an environment offering a snapshot of the lives of a number of characters as they cross paths: half-heard mobile phone conversations, incomplete journeys and the coincidences and interconnections of everyday life.

The stylistic choices made by designers do not exist within an ongoing critical narrative solely determined by aesthetic concerns. The cultural context within which such sites exist must also be considered. As author and designer Tibor Kalman once observed, alternative perspectives include the relationship of graphic design to its audience: that '…tells how political images have been crafted, how corporations have manipulated public perceptions, how myths have been created by advertising. This other history is the history of design as a medium and as a multiplicity of languages speaking to a multiplicity of people.' [19]

These new stylistic explorations reveal the human imprint of the designer on the medium – creating a relationship with the viewer based not solely on a relentless search for technical innovation and excess, but instead reflecting the tactile, tangible and often imperfect elements of the world around us.

66–93
BRANDING THE WEB
IN 1998 EVERYONE WANTED A WEBSITE, EVEN IF THEY DIDN'T KNOW WHY.

NATHAN LAUDER, FIBRE

During the mid-1990s dot.capitalism and web design became hot property. Between 1996 and 2000, companies jumped on the website goldrush, excited by the possibilities offered in the 'brave new world' of the Internet and the seemingly endless lucrative potential of e-commerce. As Nathan Lauder from design agency Fibre has said: 'In 1998 everyone wanted a website, even if they didn't know why.' However, the economic downturn in the global digital technology market in 2001 created a fundamental shift in the aims and expectations of companies looking to use interactive media as a marketing tool.

Since that time, the web design industry has seen a process of recovery and, alongside this, designers are discovering that relationships with clients and consumers are changing. Tokyo-based Business Architects sum up these changing ideals on their website: 'Now the "Net" itself stands for nothing. The once raved values of "being digital" are disappearing. No one craves for the "e" in electronic any longer. Probably it's time to use it in the word "Emotional".' [1] In addition, the audience for online content continues to grow. Jupiter Research estimates that as of April 2003, 553 million people have access to the Internet worldwide. [2]

Companies are aware of the vast potential audience for products and services and the range of choices they have in engaging with new technologies – from promotional online initiatives and viral marketing campaigns to in-store interactive signage, kiosks and temporary displays. In the UK, three-quarters of retailers are embracing digital technologies in some form. [3] However, expectations on the parts of both client and design agency are becoming more realistic about the role and value of digital media. Antti Hinkula from Finnish design agency Syrup Helsinki describes the current position, in contrast to the excitement witnessed at the end of the 1990s, as '…much more reasonable behaviour in every way. Designers are much more familiar with digital media. Clients have seen the first results and responses from consumers; they can see the importance of the web in their business. Expectations are much more realistic than a couple of years ago.'

The function of the website as part of an overall company brand strategy has become an established element in the standard design brief alongside the letterhead and the logo. The 'corporate website' can exist as a sales point, promotional opportunity, communication device or brand-enhancement tool. As Nathan Lauder says, 'In 1998 the concept of a website was exciting and full of possibilities for companies – they wanted to use the design of their website as the spearhead for the rest of the branding of the company. Now this balance has shifted so that the website is part of an overall branding and marketing package.'

At the same time, the consumer has become more brand literate – both online and offline. According to Daljit Singh, creative director of London-based interactive agency Digit, companies must become aware of the need to communicate to the public in new ways: 'Consumers are becoming more savvy. Companies need to keep up with the consumer and to keep searching for surprise elements – whether through an exhibition space, a building or a great television ad.'

For a company like Beck's, the Internet can offer a chance to create a humorous twist on their usual marketing approach. Their advertising campaign in online magazine F Magazine, for example, shows a Beck's bottle 'beaming' across the page as if from 'Star Trek'.

In a marketplace of increasing competition, companies are focusing more intensely than ever on forming deeper brand relationships with consumers. In an echo of the current trend for designers and artists to explore the creation of personal expressions online, businesses are forming new marketing strategies for digital media increasingly aimed at reaching the consumer in more experiential ways, creating two-way relationships between the brand and the user.

Levi's Europe digital marketing manager Helene Venge describes the company's growing commitment to digital marketing – through a series of websites, email marketing campaigns and an SMS strategy – as an investment on a 'more emotional level'. Venge says that the Levi's Europe website '…is taking a step from product promotion to a conceptual brand experience. Interactivity is key'. [4]

CYBER ORCHID
Shiseido Cyber Orchid Greeting Card System
Designer: John Maeda
Client: Shiseido Co., Ltd
1996

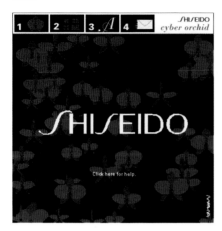
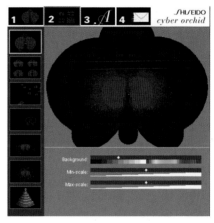

A number of companies recognized the potential for digital media to open up new connections with their customers from the early days of the web. Shiseido, for example, commissioned a series of online works by John Maeda from 1996 onwards. In conjunction with the launch of Shiseido's online orchid-ordering service, creative director Michio Iwaki invited Maeda to design an orchid-themed card-making system for Shiseido's website. Maeda created a series of single Java applets that enabled the viewer to change the flower's shape and colour, and to include personal illustrations drawn by the sender. This early e-card could then be emailed. Since 1996, Maeda has had an on-going relationship with Shiseido, creating further series of online games, cards and calendars.

Corporate marketing initiatives to encourage new kinds of relationships between the consumers and their brands are of course not limited to the Internet. Offline strategies may include the support and initiation of live events, exhibitions, media programming and other more content-led promotional activity. A growing number of corporate marketing strategies are embracing the notion of 'brand physicality'. British creative company Odd has worked with clients as varied as mobile phone manufacturer Siemens Xelibri, BBC Radio 4 and Sony Playstation to create tangible media and three-dimensional brand environments – from a mobile English country garden to a range of experiential 'zones' at Playstation Experience 2003 with names such as 'Cartoon', '18+ Boudoir', 'Fight Alley', 'Adult Creche' and 'Adrenalin'.

DIESEL FRENDSHIP GALLERY – WWW.DIESEL.COM/FRIENDSHIP GALLERY
Concept, design and curation: Diesel Creative Team
Contributing Artist/ Designer (Freedom): Rinzen
Client: Diesel
2002

Italian clothing company Diesel supports many fashion, art, music and new media projects both online and offline. In the field of new media, Diesel have embraced digital arts projects, are engaged with an ongoing collaborative relationship with Sony Playstation and have built a series of marketing initiatives that are creating dialogues between online and offline media.

For Bob Shevlin, Diesel's Director of New Media, initiatives such as the **FRIENDSHIP GALLERY** (launched in April 2002) provide the opportunity to support the work of young and upcoming creative practitioners. The Friendship Gallery also offered Diesel the opportunity to play around with and gently poke fun at the current marketing initiatives offering emotional branding experiences. According to Bob Shevlin, 'Diesel decided to take ownership of all positive emotions. Therefore whenever anyone experienced a positive emotion, it would be a Diesel Moment, brought to you by Diesel, with Diesel explicitly claiming ownership of the emotion itself.'

The Friendship Gallery campaign takes the visitor on a surreal, exaggerated and artificial journey around an online 'Happy Valley'. Within this 'Diesel Emotions Theme Park' visitors experience work by a global range of digital artists and designers who were invited to create a personal expression of their chosen emotion, including LOBO, InsertSilence, Nando Costa, NEW, Mongrel Associates, Joshua Davis and Futurefarmers.

WWW.PROTOKID.COM
Design Agency: Bloc
Designers: Rick Palmer, John Denton,
Eleanor Wilson
Programmers: Steve Hayes,
Eleanor Wilson, Jan Fex
Client: Diesel Kids S.rl
2003–4

While the Friendship Gallery was created as an online experience, Diesel have been increasingly interested in embracing campaigns that can work successfully using both online and offline channels. Diesel's **PROTOKID** site, created by the Diesel Virtual Department and developed in collaboration with design agency Bloc Media, exists as a fun online playground where visitors can customize their own 'Protokid' character, and become a famous pop star. The site was designed as a unique stand-alone communication that could continue to evolve over time, unlike most print-based campaigns which are more static in nature. However, although initially conceived as an online project, there are now increasing cross-overs with the offline world. As Bob Shevlin explains: 'The Protokids have appeared in multiple Diesel Kids catalogues, have been used in stores and displays, and now have various merchandizing items such as stickers, T-shirts, lunchboxes and bags.' For Diesel, Protokid exists as an example of a new media-driven project that is expanding into other environments.

Diesel are continuing to explore the possibilities for linking online and offline projects. At the 2003 Online Flash Film Festival (OFFF) in Barcelona, Diesel created an interactive entertaining experience by taking the theme of a recent print ad campaign and turning the entrance to the event into a James Bond-style data-processing centre, using touch-screen interactives and video animations to 'compile market research data on all Diesel individuals from around the world'.

Networked digital media initiatives have offered companies some of the most creative solutions to attracting users emotively and experientially to their products and services. The beginning of the twenty-first century has seen a growing number of high-profile corporate initiatives adopting a strategic approach to the online promotion of their brand. Businesses are investing in digital media strategies that emphasize notions of user exploration, experiential content and interactive narrative possibilities, offering brand 'added value'. Here we explore the ways in which companies can use interactive design to nurture relationships between consumers and brands, and the implications that the multimedia environment hold for the future of 'the brand'.

WEB SPACE/
AD SPACE

'Every space has become an ad space. Everything from sports arenas to nature walks and kazoo players are sponsored, branded and bartered. You're soaking in it. Blasted, assaulted, insulted, surrounded, distracted, hypnotized, and occasionally seduced by countless messages each day.'
Steve Hayden, Vice Chair of Ogilvy & Mather Worldwide [5]

The advertising ecosystem is overloaded. In 2002, every hour of daytime network TV in the US contained nearly 21 minutes of commercials. Some cable networks have 60 seconds of ads for every 140 seconds of programming. As much as 30 per cent of emails are pure spam. In the centre of this media noise the value of the advertisement is being eroded. According to Nielsen Media Research, a viewer's ability to recall a given message drops 45 per cent when the number of adverts in a commercial break doubles. Advertising has the most impact in countries with low clutter – Belgium, Denmark and the Netherlands – and the least impact in the most saturated nations: Japan, Hong Kong, Italy and Spain. [6]

For Steve Hayden, Vice Chair of Ogilvy & Mather Worldwide, the solution to this cacophony of messages is to create better adverts, ones that are so good that they become part of the culture around us, that 'resonate far beyond their original intent'. Martin Raymond, editor of trend and consumer lifestyle journal 'Viewpoint' and co-founder of futures consultancy Future Laboratory, believes that traditional marketing techniques should be abandoned. Consumers suffering from brand fatigue, termed 'brand blur' by Raymond, are no longer responding to 'beneath-the-radar methods of selling and brand building'. Instead, he recommends that marketing strategies should focus on 'building brands, products, companies that people are happy to endorse'. [7]

GIFT
Agency: The Viral Factory
Client: MTV
2001

www.mtv.co.uk/xmas At the same time, the Internet is being recognized by cultural commentators and technologists as a site of intense social negotiation. From instant messaging and weblogging to chat rooms, peer-to-peer networks and viral emails, users (dubbed the 'Generation ICQ') are embracing the social aspects of the web. For many critics these types of virtual interaction are driving the future of technology and its role in society. Lisa Strand, chief analyst of Nielsen/Net Ratings, argues that people are using a rising number of different means to socialize online: 'sites targeted towards socialization – chat, greeting cards, matchmaking services – are growing faster than the overall growth rate of the web'.[8]

The short-lived but intense phenomenon of the viral email is one element of these online social interactions. In 2002, high-profile examples such as **TABLE TENNIS** and the **STAR WARS KID** series saw a variety of different versions circulating through emails, downloads and weblogs. For a growing number of companies, creating their own branded versions of popular downloads allows them to 'piggyback' their brand into online circulation. Following the global success of the Star Wars Kid series, both Midway Games' **MORTAL KOMBAT** and entertainment sites such as **eBAUMSWORLD** created their own branded versions, while MTV created a Star Wars themed spoof for their 2001 online Christmas campaign.

For many companies, targeting the online youth market is an ongoing strategic objective. Software giant Microsoft is on the leading edge of these efforts. According to Tammy Savage, group manager of Microsoft's NetGen division, 'This [youth market] customer wants to socialize instead of communicate. They want to do things together and get things done – and they really want to meet new people. They have a way of vouching for each other as friends, figuring out who to trust and not trust.'[9]

Online branding initiatives share many of the same starting points as brand communications across all media. For the business community, the challenge comes from the attempt to translate ongoing branding efforts into an interactive experience. A successful online branding campaign requires both a relevance to the user and to be able to harness the power of the interactive medium in a way that would not be possible in other formats.

The 'brand' lies at the heart of a corporation's identity. The power of a brand derives from a curious mix of how it performs functionally and what it stands for emotionally. To be successful a brand has to have a set of rational characteristics and, as Wally Olins, co-founder of brand consultancy Wolff Olins, describes it: 'a personality that charms and seduces'. Olins continues: 'it is a mix of emotional factors, such as "Do I like it?" and "Is it me?" and rational factors such as, "Is it cheaper or better or quicker?". Getting things in balance is tricky, and that's why so many brands don't succeed.' [10]

These rational and emotional qualities have traditionally been expressed through visual elements of the corporate design – trademarks, logotypes and colour schemes. However, in moving image and multimedia environments, new constraints as well as new opportunities present themselves.

Rather than selecting a set of pantone colours for a range of print-based communications, the growth in the use of moving image promotional devices requires designers to devise a consistent 'colour style' that can exist across both static print-based media and moving image work. Creating a recognizable 'sound environment' for a brand needs to be considered, as do the possibilities for interactivity in communication. For Ken Frakes, managing director of Arnold Interactive, redefining the role of a brand online is not 'about checking that the logo colour palette is web-safe'. Instead it is important to '…step back and identify the brand's essence and role'. [11] As Austrian designer Markus Hanzer explains: 'brands can't just put themselves out there and say, "Hi, here I am". They will instead have to enter into a dialogue. A dialogue means always a certain readiness to change in respect to the one that is being addressed.'[12]

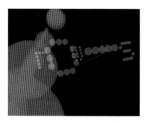

SONY CONNECTED IDENTITY
Design Agency: Tomato Interactive
Client: Sony Entertainment Ltd
2001

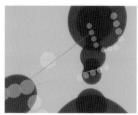

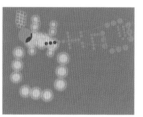

The concept of freeing up the visual identity of corporate design to dialogue and change sits at odds with the traditional idea that these elements form part of the projection of how a company would like to be perceived by its customers and must therefore exist as a construction to be cautiously cultivated and carefully protected. The **SONY CONNECTED IDENTITY** created by Tomato Interactive is one example where a corporation has not been afraid to engage in this dialogue.

With a brief from the Japanese corporation to make an identity that expressed Sony's commitment to connectivity, Tomato Interactive created a real-time living identity online that listened to and reacted to the world around it. The identity consisted of an algorithm that generated a graphic language using colour, speed, depth, height, rotation, zoom, skew and perspective to continuously update itself. For Tomato Interactive, the Connected Identity brought 'the Sony brand into the hands of the people', representing a true democratization of a corporate identity as website visitors were able to alter the identity's appearance and behaviour. [13]

Companies have a choice when creating an online marketing strategy: to dovetail online efforts with ongoing branding commitments in other media, or to use digital technology as a means to communicate the brand and its values to a specific, targeted audience. The web, alongside other media such as cable and video-on-demand, provides a channel for narrowcasting. Unlike the mass broadcast techniques of television, the web provides a site where companies can promote their product or services to a specific demographic segment – for example the youth market or the influential technologically literate 'Prosumer'. For multifaceted companies such as Levi's or Nike that offer differentiated products aimed at a range of consumers, using the web strategically as a narrowcasting device can enhance an overall marketing strategy.

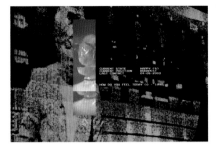

WWW.MASSIVEATTACK.COM
Design Agency: Hi-Res!
Creative Directors: Alexandra Jugovic,
Florian Schmitt
Designers: Alexandra Jugovic, Florian Schmitt,
Erik Jarlsson
Rich Media Developer: Andreas Müller
Programmer: Bela Spahn
Sound Design: Clifford Gilberto, Massive Attack
Clients: Virgin Records, The West, UK
2003

Nike have based their online strategy around a family of sites that are targeted at a range of demographic subgroups – from gender, age, language and sports interest. Excluding individual country websites, **NIKE.COM** has links to more than fifteen sites including **NIKELAB.COM**, **NIKEID.COM**, **NIKETOWN.COM**, **NIKEGODDESS.COM**, **BOWERMAN.NIKE.COM**, **NIKEPLAY.COM** and **NIKEBIZ.COM**. These sites exist as online stores, information points, play spaces and online community hotspots for athletes and sports enthusiasts.

In 2003 design agency Hi-ReS! created the online counterpart to Massive Attack's album, '100th Window'. The website sat at the heart of a wide-ranging promotional campaign that included print and moving image-based advertising, press photography, packaging and a series of music videos. Commissioned to produce a site that would extend the musical experience of the album and create an innovative community structure, Hi-ReS! created a dynamic interactive system based on the delivery and representation of data relating to the band and to the outside world.

MASSIVE ATTACK STAGE SHOW
Stage Set and Visual Design: UVA
Clients: Massive Attack, The West, UK
2003
(see also pp.78–9)

The project was designed as two sites. The first consisted of a series of flat pages presenting the core information required by the band: tour dates and band news, for example. Hi-ReS! then manipulated the same data to create something completely different with it. This took the form of a second set of pages that sat underneath the flat site: '…the site was always meant to have two sides, innocence and experience, raw accessible data and the bastardized version of the same data, mixed with [external] data-feeds'.

Hi-ReS!' creation of an underlying system of recycled data, while designed initially just for the website, has subsequently fed into other parts of the '100th Window' project. Moving image designers UVA have incorporated recycled information streams into part of Massive Attack's touring stage show, with large-scale projections of real-time data that feature comments from visitors to the website alongside other captured elements of information, creating a constant loop between online and offline, inputs and outputs.

COLETTE TRADING CARDS
COLETTELOVES

3. PHILIPPE STARCK

COLETTE TRADING CARDS
COLETTELOVES

1. KARL LAGERFELD

COLETTE TRADING CARDS
COLETTELOVES

2. GISELE

i ♥ colette

♥ DOWNLOAD COLETTELOVE!

WWW. COLETTE.FR

213 RUE SAINT-HONORÉ 75001 PARIS

Colette loves...

styledesignartfood   J'TAIME COLETTE

DESIGNED WITH LOVE BY FLIP FLOP FLYIN'

WWW.COLETTE.FR
Design Agency: Spill
Client: Colette
2002

'COLETTE LOVES' TRADING CARDS
Designer: Craig Robinson
Client: Colette
2000

WWW.ILOVECOLETTE.COM
Designer: Craig Robinson
Client: Colette
2000

Websites can also be used as a counterpoint to an existing marketing campaign. Design agency Spill created **WWW.COLETTE.FR** for the fashionable Parisian 'concept store' Colette, using illustrated backdrops of the boutique and cheeky French canines as navigational devices. In 2000, Colette invited British designer Craig Robinson to design an alternative site, **WWW.ILOVECOLETTE.COM** , which reinterpreted the 'official website' created by Spill and translated into his own unique style. www.ilovecolette.com featured a number of motifs from the Spill-designed website, including using hand-drawn versions of the shop as an organizational device, but added additional elements that included a set of online 'Colette Loves' trading cards featuring Robinson's signature pixelated miniature portraits.

THE BRAND AS EXPERIENCE: SONY AND NIKE GO ONLINE

At the heart of many corporate online marketing strategies is the pursuit of new ways of creating connections between consumers and the brand: through brand-led content development and collaborative ventures with artists and designers, or through the creation of interactive narrative structures and immersive environments. The most successful initiatives are ones that recognize and are able to engage with the transformative, interactive, engaging nature of the online environment.

Several recent projects have been initiated by commercial clients that showcase the work of leading web designers and artists using a collaborative framework – reminiscent of self-initiated online collective projects such as the **REMEDI PROJECT** or the **CODEX SERIES**. These collaborations provide a means for companies to engage with the viewer in a more emotive and experiential way. In addition to Diesel's Friendship Gallery (see page 70), other online art and design-led projects have included the **MIND THE BANNER** initiative by Japanese telecommunications corporation NTT, and award-winning websites by Sony Playstation and Nike. These projects provide a way for businesses to target the tricky tech-literate teen and youth markets, and to support the work of creative practitioners; they can provide a way for companies to explore the potential of 'branded content'.

While some designers have raised concerns about the increasingly commercial face of creativity on the web, designer James Tindall thinks that corporate-sponsored web art sites can be a positive development. As Tindall explains, 'because of time constraints and pressures to concentrate on corporate work – personal sites are often superbly built and beautifully packaged, but rarely anything beyond the glossy facade'. Sponsored brand-building exercises 'offer digital artists and designers one possible solution to funding problems along with enough creative freedom to allow further development of their ideas'. [14]

SONY PLAYSTATION 2
Online advertisement campaign
Director: David Lynch
Advertising Agency: TBWA/London
Client: Sony Entertainment Ltd
2000

WWW.THETHIRDPLACE.COM
Design Agency: Hi-ReS!
Creative Directors: Alexandra Jugovic,
Florian Schmitt
Designers: Alexandra Jugovic,
Florian Schmitt, Erik Jarlsson
Shockwave Programmers:
Andreas Müller, Simeon Rice
Programmer: Bela Spahn
Sound Design: Clifford Gilberto
Client: Sony Computer
Entertainment Europe
2002–

In 2002 Sony launched **WWW.THETHIRDPLACE.COM**, an online space featuring new interactive work by a series of well-known web designers and artists curated by design agency Hi-ReS! This website formed one element of a wider brand context that Sony had developed for the Playstation 2 games console, titled the **THIRD PLACE**. From the marketing campaign that launched the Playstation 2 in 2000, Sony have chosen to downplay the technology of gaming – removing images of the console or games from television, cinema, billboard and magazine adverts – and have instead focused attention on the environments and experiences it has enabled game creators to produce.

84–85

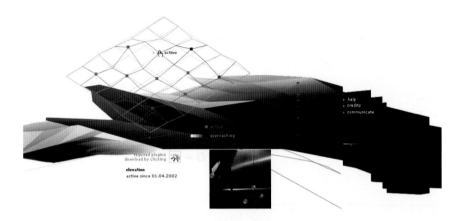

Inspired in part by Ray Oldenburg's theory of third places introduced in his book 'The Great Good Place', Sony have defined the Third Place gaming environment as a surreal, dreamlike space where simulated adventures are promoted ahead of physical experience. In the black-and-white cinema trailer directed by David Lynch that launched the Playstation 2 console in the UK, Sony introduced this sinister, dreamlike space, characterized by abrupt and disturbing transformations to the environment and its occupants. This sense of dislocation was heightened in following adverts with the insertion of unsettling glimpses of the gaming environment into mundane real-world situations.

Like 3D virtual communities, this presentation of the Sony Playstation 2 gaming environment holds out the promise of satisfying many of the desires often associated with virtual environments: freedom from the confines of the physical body, freedom from the constraints of geographical space, freedom from the fear of contact with others and freedom from the ordering structures and restrictions of society. Like Oldenburg's vision, this virtual space is a neutral, public space offering some form of social interaction, if from a distance; however, Sony's Third Place offers an escape from place altogether, releasing the virtual visitor from the constraints of the physical world and the responsibilities associated with a fixed identity.

WWW.THETHIRDPLACE.COM was launched as one element of this wider brand context for Sony's gaming environment. The site enabled Sony to create an immersive online environment that would encourage visitors to explore an interactive embodiment of the Third Place concept. At the same time, the website presented an experimental showcase for some of the most respected web designers currently working, including Daniel Brown, Joshua Davis and Yugo Nakamura. Hi-ReS! created a 3D environment that featured a series of floating interactive works that could be entered by visitors to the site. The brief that the participants received from Sony was straightforward: simply to present their vision of 'the third place' using their own design aesthetic.

As Florian Schmitt describes: 'the beauty for us was to be able to present all these very different people, all with their different motivations and cultural backgrounds, to the wide audience that Playstation provides'. For Schmitt, the website benefited greatly from this collaborative framework, a way of approaching projects that for him typifies much of the work being created both commercially and non-commercially on the web: 'Many people put their work online, they collaborate, they get things moving, and that's one of the things that we have always loved about the web, its utter democracy and ubiquity.' [15]

The Nike Swoosh is everywhere: for better or worse the Nike brand has become embedded in the global consciousness. Part of this success has undeniably been created through a series of memorable and award-winning advertising campaigns. In the view of Dan Wieden, co-founder of Nike agency Wieden + Kennedy, part of this success has come from a willingness to take risks: 'there's a mutual desire to experiment – it's not so much about trust as "let's find out"'. This spontaneous approach suits Nike co-founder Phil Knight, who says of Nike's advertising approach: 'Our job is to wake up the consumers.' [16]

WWW.NIKELAB.COM
Design Agency: R/GA
Commissioned Designer: Yugo Nakamura
2002

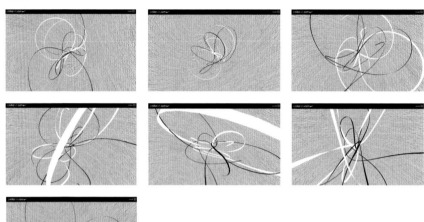

Nike's experimental approach to building new promotional strategies that reach a broad span of their consumer market includes a strategic approach to the role digital media can play. The Internet not only provides a key medium for reaching targeted groups of consumers, but can also provide companies with a freedom to explore and take risks with promotional campaigns that would not be seen as appropriate for other mainstream broadcast channels. For Nike, an example of this came in 2000 when, as a television spot, an advert featuring an Olympic runner being chased by a lunatic with a chainsaw was pulled by US channel NBC after receiving viewer complaints; however, it fared better online as an extremely popular web download.

In August 2001 New York based R/GA (formerly R/Greenberg Associates) was awarded the Nike US Interactive Agency account. Initially founded by Bob and Richard Greenberg in 1977 as a leading-edge motion graphics and live action film and video production company, R/GA has become a multiplatform interactive advertising agency with a portfolio that spans a wide range of digital projects. R/GA created the sites and online marketing initiatives for a number of Nike websites before working with Wieden + Kennedy on the concept for **NIKELAB** in 2002.

WWW.NIKELAB.COM
Design Agency: R/GA
2002

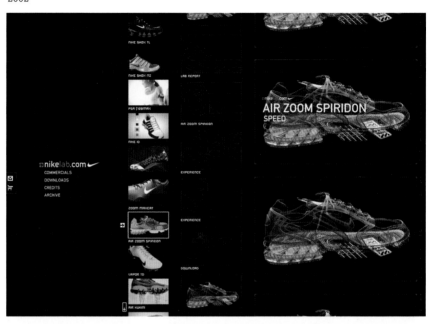

WWW.NIKELAB.COM
Design Agency: R/GA
2002

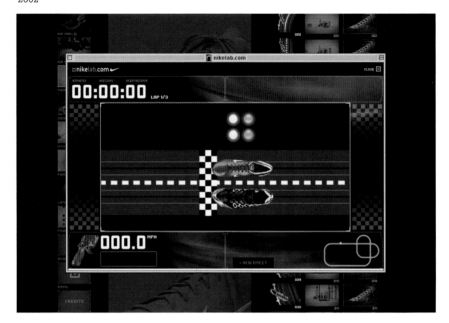

Given the brief to 'communicate Nike's innovation and inspiration in a cool and fresh way', Rei Inamoto, creative director at R/GA, worked closely with Wieden + Kennedy and the Nike team to draw up a concept based around the framework of a series of artist collaborations. The project could have evolved into a DVD, a kiosk or any other digital medium available, but in the end Nikelab became an online destination. For Inamoto this was the best solution for a number of reasons: 'it is an extremely efficient way to reach a large audience; a website is a flexible medium allowing content to be added over time, and the concept behind the site was something that had not previously been done'.

At the heart of the Nikelab project for Rei Inamoto was the desire to engage the visitor: 'In order to truly communicate Nike's innovation and inspiration, we quickly realized that literally saying "Nike is an innovative and inspirational company" was not the way to go. A more effective way to communicate that message was to create an experience that was innovative and inspirational in its own right and would evoke a visitor's visceral and emotional response.'

WWW.NIKELAB.COM
Design Agency: R/GA
Commissioned Designer: Edwards Churcher
2002

R/GA acted as a 'digital art curator', commissioning web designers and artists to create new work related to individual Nike products. Each product had a primary keyword associated with it, and each artist was asked to express that keyword in some way. Mercurial Vapor, an ultra-lightweight football shoe, was associated with the word 'speed'. The work produced by Edwards Churcher and other Nikelab artists featured interactive metaphors for 'speed', including a slot-car racing game.

R/GA produced a continuously evolving immersive cinematic experience, creating a structure inspired from moving image production, architectural practice and exhibition design rather than print-based graphic work. As Inamoto explains: 'the team wanted Nikelab to be an "experience", so we were careful not to try to recreate, say, a "room" that displayed products and works by artists…It was taking the thinking behind [architecture and exhibition design] disciplines and how individuals interact in a space that helped us create a completely different online experience for Nikelab.'

At the centre of Nike's promotional strategy is a growing interest in the role brand-led content can play. In addition to Nikelab, recent offline initiatives have included the broadcast in 2002 of television documentary 'Road to Paris', a joint production between Nike, Wieden + Kennedy and production company @radicalmedia that followed Lance Armstrong's bid to win the 2001 Tour de France. Nike intends to increase its participation in future documentaries, with proposed films including profiles of Michael Jordan and Roy Jones Junior.

THE
BRAND AS
CONTENT

Brand-funded content is of course not new – examples of the attachment of positive associations to brands through product placement, product endorsement and sponsorship of entertainment programming are all around us. Increasingly, however, the trend is for brands not just to attach themselves to entertainment, but to create their own. The growth in digital TV and broadband, and the proliferation in media channels, have opened new opportunities for advertisers to look beyond the thirty-second commercial break. It has been estimated that by 2005, 25 per cent of advertising revenue will be spent on such content-led initiatives. Already, high-street retailer Boots, in conjunction with commercial television producer Granada, has launched health channel Wellbeing, and as more channels seek more content, this phenomenon is set to grow.

Nick Barham, director of cultural research unit Profusion, feels that brand-led content can provide the consumer with the entertainment that they want: 'they offer a deeper experience. You get more time with the brand but, more importantly, these are things that people will choose to experience.' [17]

Interactive media, and the Internet in particular, can provide a medium for companies to explore more content-driven promotional strategies in two key ways. With the rise in broadband capabilities, advertisers are looking to the Internet to provide a distribution channel for branded content that can be targeted at specific audiences, is inexpensive and allows repeated hits by visitors.

'The Hire' is a series of action-packed short films developed by BMW in association with Fallon Worldwide and director David Fincher. Available from **WWW.BMWFILMS.COM**, 'The Hire' is a prime example of a brand creating entertaining content for a tightly targeted audience. Achieving more than thirteen million hits by the end of 2002, the series features high production values: stunning car chases, a star-studded roster of directors including Tony Scott, Ang Lee, John Woo, John Frankenheimer and Guy Ritchie, and acting performances from F. Murray Abraham, Gary Oldman, James Brown and Clive Owen.

While this initiative by BMW uses the Internet as a flexible distribution channel rather than for its interactive qualities, other companies are looking at the potential for the web to support branded content in a much more responsive, immersive way. Both Diesel's Protokid website and the Minority Report Experience by Lexus offer creative interactive narrative structures that are engaging and fun, but also keep the user in contact with the brand far longer than a static presentation would.

WWW.HABBOHOTEL.COM
Designer: Sulake Labs
2000

Advertisers are also spotting the potential for targeting specific audiences through the social channels of the Net, from viral emails to chat rooms. Winner of a Golden Nica for Net Excellence at the 2003 Prix Ars Electronica, **WWW.HABBOHOTEL.COM** is one of the most popular sites on the Internet. Launched in January 2001, Habbo Hotel exists as a virtual hotel where you can explore and make new friends among the growing community of more than three million members. Members create their own customized animated character, known as a 'Habbo', who can walk, dance, eat, drink and chat in the various online spaces. For Sulake, the game development company behind Habbo Hotel, the site also offers opportunities for alliances with third-party brands to 'reach the highly desired twelve–eighteen-year-old market in a cost-effective and creative manner' by offering a platform for product placements and sales pitches. [18]

As the search for the 'brand experience' becomes the holy grail for advertisers and their corporate customers, the interactive environment will continue to offer a space for creative partnerships, brand-led content and immersive user experiences. As digital media channels converge with broadcast production, the Internet, mobile technologies and wireless networks, the future challenge for businesses will be to harness the potential for these 'brand experiences' to be recreated across multiple platforms, providing a responsive, holistic approach to brand promotion.

NOKIA GAME
2003

The **NOKIA GAME** series may signal where future multimedia marketing strategies may be heading. Nokia Game is a 'ten-day worldwide gaming experience' where players from eighteen countries can complete challenges and solve clues across a number of different media platforms – from mobile networks to the Internet, TV, radio, magazines, voicemail and press ads. [19] To complete the game, participants need to stay connected via the **WWW.NOKIAGAME.COM** website for the entire length of the event. This marketing tool drives users to the Nokia website, but is also entertaining, creates a feeling of community between gamers and is able to reinforce Nokia's 'connecting people' brand values.

At the same time that companies are embracing the opportunities offered by new technologies, a number of cultural commentators are raising concerns. Writers like John Seabrook, a critic-at-large for 'The New Yorker' and author of the influential book 'Nobrow: The Culture of Marketing, the Marketing of Culture', have voiced concern over the situation where it is the commercialization of culture that confers status and where content and advertising sit side by side in what Seabrook terms 'the Megastore' of contemporary society. For Seabrook, these attitudes draw attention away from the 'soul' of corporations and the almost forgotten values of quality, consistency, integrity and honesty. [20]

While the birth of the Internet brought with it a critical dialogue based around the notions of free access, community action and personal involvement, the 'free' nature of online environment also encourages this cross-breeding between editorial content and advertising, high and popular culture, news and entertainment. As Rick Poyner explains in his book 'Obey the Giant: Life in the Image World', there are no pre-existing protocols to determine how editorial and advertising terms are assigned online. [21] From a space that seemed to offer the viewer an escape from the structures of everyday society, the Internet now provides an opportunity for 'the brand' to become the cultural infrastructure itself, no longer simply an add-on or interruption. For designers, clients and consumers, future initiatives will need to engage not only with technological advances, but also with the changing cultural and social trends that accompany them.

FUTURE
INTERACTIONS

DESIGNERS ARE LOOKING BEYOND THE SCREEN CULTURE TOWARDS MORE TACTILE, TANGIBLE EXPERIENCES, CREATING NEW KINDS OF DIRECT CONNECTION BETWEEN PEOPLE AND TECHNOLOGY.

Two key developments are influencing the future of interactive design. Firstly, continuing technological innovation will offer new communication channels for designers. Perhaps more significantly, however, designers are increasingly focusing on the social, economic and political implications of people's relationships with technology.

In the past decade the convergence of computing with telecommunications, in particular mobile communication and the Internet, has increased the intensity of electronically transmitted connections in our lives. Boundaries are blurring between information and entertainment, work and leisure, public and private. The ways that individuals and society view themselves and their relationships are increasingly being filtered through these digital nodes and networks, and among these intertwining strands designers are looking to engender more personal, emotional and experiential relationships.

A growing number of institutions are advocating this approach: balancing the aesthetics of interactive design within a contextual critical framework. The highly regarded Interaction Design Department at the Royal College of Art in London and MIT's Media Laboratory have been joined by two recently founded European institutions: the Interaction Institute in Ivrea, Italy, and Hyper Island School of New Media Design, established in Karlskrona, Sweden in 1996.

Each of the courses offered by these institutions encourages students to consider not only the commercial applications of projects, but also the human context of technology, as Irene McAra-McWilliam, head of the Interaction Design Department at the RCA explains: 'It isn't enough simply to understand the possibilities offered by new technologies. Interaction designers also have to understand people: how they experience things, how they themselves interact and how they learn. It's not about the design of the next microwave; it's about the role of the kitchen.'[1]

BIOPRESENCE
Human DNA Trees
Designers: Georg Tremmel
and Shiho Fukuhara
2003

WAITING TIME
Part of SLOW TIME project
Designer: Megumi Fujikawa
2003

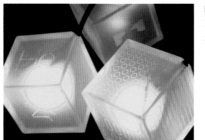

LOOPQOOB
Designer: Murat Konar
2002

The student projects featured in the RCA's Interaction Design Department end of year show in 2003 reflect these contextualized concerns – featuring an electronic confessional that can absolve you by text message, a bus stop that encourages people to sing while they wait, cubes that make music as you move them and even a project that allows you to put the DNA of a deceased loved one into a tree to form a living memorial.

These students will join a workplace in the process of change. Technologies, media and creative disciplines are converging, and as this process continues the role of the designer will also change. As commercial multiplatform strategies become more prevalent, the skills required by interactive designers will evolve. Bob Greenberg, chairman and CEO of interactive advertising agency R/GA, states that designers will need to build truly multimedia skills: 'the designer of the future will need to conceptualize an idea and then repurpose that brand experience across multiple devices and channels – from narrowband, broadband and wireless to print, broadcast, radio, direct marketing and billboard advertising'. According to Greenberg, this 'cross-media expertise', along with the knowledge of changing digital and technological tools, will allow the designer to successfully implement the effective integrated and cross-platform campaigns increasingly required by brands.

The concept of technological convergence – the coming together of narrowband, broadband and wireless communications – has been in the public domain for some time. For critics, the slow emergence of interactive television (iTV), the disappointing performance of mobile communication protocol WAP and the underwhelming public reaction so far (at least in the UK) to 3G (third generation) bandwidth mobile phones have all been said to be indicative of the fact that the promise of digital convergence – that it would change the way that societies and individuals communicate, work, play and live – is nothing more than naive futurism. Two key events have, however, made the promise of convergence seem more realistic than ever before: the rapid growth in high-speed Internet connectivity and the recognition of the potential of wireless networking and mobile communication by major global corporations.

In July 2003 the EU introduced new laws bringing together all forms of electronic communication legislation – wireline, wireless, cable and satellite. [2] In the same way that digital technology is erasing the distinctions between voice, data and video, the EU is changing its laws to follow suit. The development of global digital standards alongside legislation will enable companies to take advantage of converging digital technologies, create consistent visual messaging across multiple platforms and provide exciting opportunities for designers to create innovative interfaces for commerce and communication.

The gap between domestic-use broadband and narrowband is closing, as Internet users reject dial-up models for high-speed Internet access. Research suggests that while 27 per cent of all US homes currently use broadband connections, by 2008 this figure will have grown to more than 70 per cent, or approximately 64 million subscribers. [3] 'Despite the slow economy, consumer demand for broadband was remarkably strong in 2002, when the US market grew by more than six million subscribers,' notes James Penhune, director of Strategy Analytics Global Broadband Practice. 'Over the next five years, high-speed access will become the norm for residential Internet users as broadband becomes more widely available, more flexibly priced and a more powerful vehicle for new kinds of entertainment, content and services.' [4] The take-up of broadband Internet connectivity is accelerating even more rapidly in Japan and Europe.

These new services can provide enhanced opportunities for video-on-demand services (such as Video Network's HomeChoice service in the UK), multiplayer online gaming and streaming media pick-up – including online 'television' and 'radio' channels. New broadband capabilities are enabling the Internet to exist as a viable alternative distribution channel for high bandwidth, rich media experiences, including, for example, the series of 'The Hire' online films from BMW (see page 91).

Companies are beginning to look to the web to provide high-quality streaming opportunities for live events – for example regular showcases of live sporting events, as in the case of Nike site **WHATEVER.NIKE.COM.** Several broadcasters are experimenting with ways to enhance digital and terrestrial television programming with simultaneous Internet streaming (known as 'two-screen' convergence), or dual programming for TV and PC. While broadband access does not alter the innate qualities of the Internet, it can provide users with enhanced media experiences and enable media companies to consider new ways of creating consistent visual messaging and content delivery across a growing number of distribution channels – in both broadcasting and narrowcasting terms.

The convergence of handheld wireless devices, telecommunications and high bandwidth computing capabilities is offering new communication strategies for businesses and consumers. With the rise of public access wireless LAN hotspots and wireless home entertainment networks, and the introduction of 3G phones, the future of technology is being sold to the consumer as a future of communication and connectivity. New technologies are being offered to the consumer as a way of bringing people together rather than isolating them.

These concepts are perhaps most clearly visible in the role of mobile technology in society today. For many of us, life without a mobile is scarcely imaginable – in a recent survey of British young adults, 46 per cent described the loss of their mobile phones as akin to bereavement. [5] Mobile phones can play a critical role in the nurturing and development of our most intimate relationships – our personal networks of friends, colleagues and family. Functioning as 'comfort objects', antidotes to the hostile environment of wider society, our mobile phones can seem intimately part of ourselves.

This intimate attachment comes not only from the regular physical proximity of our phones to our bodies, but also from 'the imagined connections concealed within it' – to friends and loved ones. [6] Evidence of the personal and emotional relationships that many, particularly teenagers, have with their phones can be seen in the way they delight in personalizing ring-tones, wallpapers and phone covers. In 2001, a MORI survey found that 3.4 million people in the UK change their mobile ring-tone at least once a month. [7]

In the same way that mobile communication can sustain existing relationships, mobile phones can also help to reinvent our ideas of the local environment. Unlike the Internet, which is often presented as a globalizing technology, mobile phones can serve as a means of connection 'any time, anywhere, any place', remapping the locality by creating the kinds of spontaneous unplanned connections available to close-knit pre-industrial communities. [8]

Network providers and product manufacturers are reflecting these changing attitudes towards mobile phone technology through the adoption of a 'soft approach' to the promotion of new mobile communication capabilities. Just a decade ago in the UK, the mobile phone was an aspirational object, presented as an expensive accessory of entrepreneurs or the well heeled. As James Harkin describes: 'In the early 1990s Nokia UK presented the mobile user as a "City Man", a smiling businessman striding through an urban landscape and wearing an expensive smile.' [9]

Today's marketing approach is very different. 3G web-enabled phones are promoted as an opportunity to communicate to friends and family through video and still image capabilities, to receive streamed sporting events or to download and play online multiplayer games. In part, this focus on the 'fun' aspects of 3G can be linked to the disappointing take-up of WAP-enabled phones at the end of the 1990s. The 'Internet in your pocket' failed to live up to expectations. 3G service providers are instead choosing to downplay the technology behind their mobile products in favour of promoting the positive effects that these devices can have on the consumer. The marketers now tell us that with our 3G phone we can have an enriched life – closer connections with friends and family and much more fun.

**FUN ANYONE? –
SONY PLAYSTATION 2**
Illustrator: Erica Åkerlund
Advertising Agency:
TBWA/London
Creatives: Frazer Jellyman,
Alasdair Graham
Animation Director: Russell
Brooke
Client: Sony Entertainment Ltd
2003

The 'soft approach' is not limited to mobile phone technology – other technology providers, particularly games manufacturers, are also choosing to focus on the fun elements of their product. For Sony's Playstation 2 this has meant a focus on the games player and the act of playing rather than the technology involved in recent campaigns. As TBWA/London art director Frazer Jelleyman describes: 'Gaming in general has become too serious, too dark and sinister; it's become something enjoyed by the few rather than the many and, of late, that's been reflected in its advertising.'[10] The animated sketches by Finnish illustrator Erica Åkerlund feature likeable characters in silly situations before the hand-drawn end line 'Fun Anyone?' pops onto the screen.

For a designer, the act of designing for mobile devices can be a very different experience from presenting content on a computer screen. Even with the emergence of 3G-enhanced technologies, mobile phones, personal digital assistants (PDAs) and other wireless handheld devices still present content in a mainly static, text-based form. As a new platform, these mobile devices currently provide limited opportunities for creative screen-based design with a restricted use of colour, basic browser technology and a typical display screen of only 128 by 128 pixels; however, despite these limitations a growing number of designers are choosing to explore the challenges of this new design frontier. For Tomato Interactive designer Tota Hasegawa, the restrictions of the medium encourage rather than discourage a creative response: 'I am most excited about the actual medium that is very limited; you just can't do very hi-tech visuals. It made us concentrate instead on strong ideas and concepts rather than just putting fluffy effects on the screen.'[11]

Japanese companies and designers have been at the forefront of experimenting with interactive design for mobile devices at the start of the twenty-first century. Several initiatives from Japan have invited designers to create interactive wallpapers, games and calendars for mobile phones. One of the first – **MOBILE GALLERY** – was created by Japanese design group Delaware between 2000 and 2001, and was followed by Japanese mobile phone art site **THE END** in 2001. Tomato Interactive, John Maeda and Toshio Iwai were among the designers invited by The End to create a Java clock or game for the NTT DoCoMo iMode mobile phone.

Design agency Syrup Helsinki received a commission in 2001 to create the visual image, branding and advertising for Finnish mobile visuals start-up company **YAKUTA**. This offered them the opportunity to look at new ways of approaching design for mobile phone devices. Designers Antti Hinkula and Teemu Suviala came up with the idea for a 'Yakuta Collection' – a series of exclusive wallpapers for new mobile phones with colour displays.

Hinkula and Suviala wanted to create a new kind of product in a very standardized market, rejecting the generic and bulk-produced wallpapers and graphics on offer commercially in favour of what they term 'a design/fashion service'. They invited a range of young designers to create monthly collections of mobile visuals that could be downloaded by customers: 'the brief given to the designers was simple – we gave dimensions of the artwork and format needed, and otherwise everybody had a totally free hand to create anything they wanted to'. Since the launch of the service a multinational group of graphic and web designers have been invited to participate, including Japanese designers Delaware, Brazilian design and animation studio Lobo and design group Abäke from the UK.

WWW.YAKUTA.COM
Design Agency: Syrup Helsinki
Art Direction, Design and Programming:
Antti Hinkula and Teemu Suviala
Client: Yakuta Ltd
2001

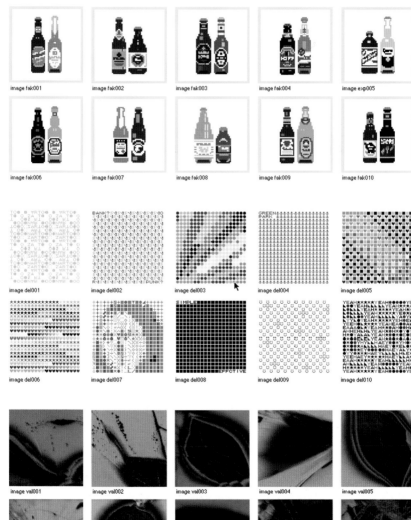

image fak001

image fak002

image fak003

image fak004

image exp005

image fak006

image fak007

image fak008

image fak009

image fak010

image del001

image del002

image del003

image del004

image del005

image del006

image del007

image del008

image del009

image del010

image val001

image val002

image val003

image val004

image val005

image valz006

image val007

image val008

image val009

image val010

For Syrup Helsinki, mobile devices offer great opportunities for designers in the near future:

> Like computer technology, memory size in mobile devices is growing in huge steps, allowing more data to be stored in your phone, more images, and bigger and better quality sound files and video. Connections are getting faster and larger chunks of information can be sent from device to device. As more phones are sold, more services will pop up for them. Services need content and content needs designers…we believe there will be a big market for designers in this environment.

For many commentators, at the heart of these converging interactive technologies lie new possibilities for gaming. Many games publishers and mobile telecommunications companies see wireless gaming as the key to unlocking the 3G mobile phone market. Walter Deffor, chief executive officer of mobile games provider Wanova, states: 'Mobile gaming is set to be the single most important driving force behind the take up of GPRS [a follow-on protocol to WAP] and 3G.' [12] On the one hand, mobile phone networks are looking for enhanced content, and on the other, the computer games industry is looking for a way to diversify away from the high-spend, high-risk games console market. For many, wireless gaming seems like the ideal solution, and mobile networks in particular are betting large sums on the success of game-playing phones. It may be that these bets are well placed: research by market research firm Datamonitor suggested that the current practically non-existent wireless game market in Europe and the US will grow to a lucrative $6 billion by 2005. [13]

Telecommunications companies are drawing together a range of multi-media strategies to take advantage of this growing market. The 'big four' operators Ericsson, Motorola, Siemens and Nokia have joined forces to build the 'Universal Mobile Games Platform', while building on-going alliances with the games industry. For Nokia, initiatives such as the **NOKIA GAME** have created opportunities for promoting their brand in this new multimedia, cross-platform environment.

These new mobile devices, however, can be much more than playthings. A key challenge for designers will be to balance the fun aspects of new technologies with an awareness of the potential that they have for wider social, political and environmental change. The real long-term value of the 3G mobile device lies in the possibilities for interaction with other people rather than with ourselves. Greater bandwidth, location-based technologies and the 'always on' nature of 3G could have a far-reaching impact on our social relationships, and offer a more consumer-centric model of public service delivery – from e-voting and access to government information via mobile devices (using Wireless Mark-up Language, or WML), to basic travel, health and education services. Transport for London (TfL), for example, included SMS as one of six payment methods for the congestion charging scheme, and during the first four weeks of the scheme 15 per cent of all payers opted to pay by text. [14]

The greatest implication for the use of mobile technologies today is the potential mobile coordination has to spur wider patterns of social change – from organized political action (such as the 1999 World Trade Organization riots in Seattle) to the intensification of the 'urban metabolism'. [15] The new generation of location-aware technologies, partnered with connectivity hotspots and widespread broadband availability, will give rise to the 'real-time city' – with instantaneous information dissemination and retrieval, richer content and peer-to-peer interactions.

FUTURE INTERFACES: BEYOND THE SCREEN

While technological innovations are offering designers new channels for creative expression, a group of designers are looking beyond the screen culture towards more tactile, tangible experiences, creating new kinds of direct connection between people and technology.

The over-arching computer-human interface paradigm upon which both Apple and Microsoft Windows operating systems are based – the computer screen featuring hierarchical file management systems and windows, icons and pointing devices – was developed almost thirty years ago at the Xerox Palo Alto Research Center in California. Screen-based interactions have given rise to new representations of space. For new media commentators, interactions with this space beyond the screen can be seen to entail a suspension of the real and physical self, or its substitution by a disembodied, virtual surrogate.

While these distancing mechanisms offer a 'virtual playground' providing unlimited opportunities to what can be imagined or acted out – like the dreamlike **THIRD PLACE** constructed by Sony – this virtual empowerment can come at a price: the loss of tangible, physical relationships with our own bodies and those of others. The growing dominance in contemporary culture of the visual over the tactile can create disembodied, distanced relationships, rather than the immediate and implicated dialogue created through touch:

Unlike the probing finger, the eye leaves its object of exploration unchanged…The body as the place of action and forceful interchange with the world for the moment fades away. This is intensified by the spatial distances sight opens, allowing the subject to dwell experientially far off. In touching, one's own body remains a proximate copresence with the touched, always immediately implicated. [16]

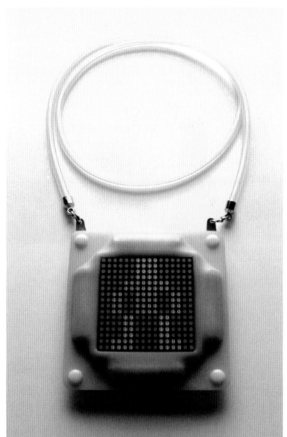

BITMAN
Designer: Ryota Kuwakubo
1998
© Maywa Denki,
Ryota Kuwakubo

Alongside the students of the Royal College of Art, the Interaction Design Institute Ivrea, Hyper Island and MIT, a growing number of designers, artists and technologists are looking to new physical interfaces and interactive environments to explore ways of bypassing the distancing mechanisms of the screen and engendering more tangible and personal connections between users and technology. The desire to explore physical connectivity within the interactive sphere is one that crosses boundaries between graphic design, product development and environmental design.

Ryota Kuwakubo is a Japanese 'device artist' who explores the dialogue between digital media and physical environments. Best known for product and software development, Kuwakubo has created portable games machines and interactive devices that respond to physical actions, including the **BITMAN** 'interactive toy' and **BLOCKJAM**, a musical interface created in collaboration with Sony CSL Interaction Lab and Sony Design Center. BlockJam's physical interface is created through the arrangement of twenty-four blocks – by positioning the blocks in different ways, musical phrases and sequences are created, allowing multiple users to play and collaborate.

MOVEMENT EXHIBITION
GMI project by Tomato Interactive
Designer: Joel Baumann
Sendai Mediatheque
2001
Photo: Fabian Monheim

The Movement exhibition at the Sendai Mediatheque in 2001, developed by Kuwakubo alongside Akira Natsume and Hitoshi Saeki, explored the relationships between graphic design, interface design and the physical environment. Ten different works were featured by a multinational group of graphic designers and artists – including Tomato Interactive, Shynola, Yugo Nakamura and Experimental Jet Set. Each piece in some way captured the movements, sounds and reactions from visitors, triggering a variety of simple electronic responses. From dancing, singing, turning the pages of a book or pulling a series of ropes, visitors' actions prompted the projection of moving images, interactive animated characters or the flipping of television or radio channels.

Artists' group Blast Theory create award-winning interactive performances and installations that use technology in the physical environment not as a means to an end, but as a way of engaging with social and political concerns. Founded in 1991 by Matt Adams and Ju Row Farr – and joined in 1994 by Nick Tandavanitj – their work embraces a range of media and disciplines, from virtual environments to video documentaries. Blast Theory mix live audience participation, interactive design elements and the use of the latest digital communication technology to create performances built around a commitment to engaging with their audiences in an immediate, participative way.

The group pioneered the use of new technologies within performance contexts with early work such as GUNMEN KILL THREE, CHEMICAL WEDDING (both 1992) and STAMPEDE (1994) – these drew on the high energy and accessibility of club culture to create multimedia performances that encouraged participation. Their use of new technologies included digital video projection, interactive pressure pad systems triggered by audience members and video and audio streaming.

CAN YOU SEE ME NOW?

Interactive game played online and on the streets using
handheld computers with GPS and GPS receivers
Artist Group: Blast Theory
Collaboration with the Mixed Reality Lab, University of
Nottingham and commissioned by Shooting Live Artists
2001

Since 1997 the group have
diversified into online, installation and new media works in participatory
performance contexts, forming an ongoing collaboration with the Mixed
Reality Lab at the University of Nottingham. Recent performances **CAN YOU
SEE ME NOW?** and **UNCLE ROY ALL AROUND YOU** have been organized around
real-time gaming situations that have featured mobile communication
devices, webcams and interacting online and offline players.

Blast Theory's work explores the relationship between real and virtual
space – using performance, video, mobile and online technologies to ask
questions about the ideologies present in the popular culture and information channels that surround us. For Matt Adams, both Can You See Me
Now? and Uncle Roy All Around You exist as part of a lineage of works
exploring the cultural resonance of the city and new technologies –
specifically mobile devices.

These pieces evolved from the notion that mobile technology is
transforming the social dynamics of the urban environment, changing the
city into a series of communication sites: 'Whereas ten years ago, you
would arrange to meet in a pub and everyone would converge there, now
everyone has mobile contact. So social arrangements are deferred and
everything becomes contingent on other things falling into place.' Equally,
it is now common to conduct conversations in a public rather than private
context, as Matt Adams explains: 'The kinds of places in which you have
certain kinds of conversations have completely broken down.' Blast
Theory's work reflects the abrupt reorientation of our traditional notions
of public and private space by mobile usage. As the mobile phone is used,
the public sphere becomes subsumed into a series of bubbles of personal,
private spaces.

Blast Theory use technology to encourage intricate interactions between people and to break down the boundaries between audience and performer. At the heart of their work is 'an interest in engaging people in a specific time and place' through what Adams terms 'the really basic concept of just communicating with each other'.

This simple notion of 'just communicating with each other' lies at the heart of the recent piece Uncle Roy All Around You (2003). Uncle Roy creates an interactive version of an espionage movie that takes place on simultaneously real and virtual London streets, using the format of a game in which you are given the task of finding Uncle Roy. Street Players use PDAs to receive maps and clues to direct them through the real city while Online Players cruise through the virtual city, listening in to the Street Players and swapping information. While the project was predicated upon the use of sophisticated technological devices, the work ends with the creation of real, tangible connections between players. The game culminates in the discovery of Uncle Roy's office. As the Street Player enters the real environment he or she makes contact with Online Players through a webcam of the office space. On their arrival Street Players and Online Players are asked to make a tangible social contract to each other. Matt Adams says that this social engagement is part of a desire by Blast Theory to engender the participants of their work with a sense of commitment to one another.

UNCLE ROY ALL AROUND YOU
Interactive game played online and on the streets of London
using handheld computers
Artist Group: Blast Theory
Collaboration with the Mixed Reality Lab, University of
Nottingham and supported by an Arts & Humanities Research
Board Innovation Award, Equator, BTexact, Microsoft
Research and Arts Council England
2003

While the commercial potential for mobile communication is being embraced by commentators, companies, marketers and designers, the role of tangible interactive environments within the commercial sphere is still to be fully investigated. Several retailers have begun to explore the benefits for them of utilizing interactive media beyond the computer screen – Levi's, for example, commissioned an interactive kiosk by Antirom as early as 1997, while other projects have been commissioned by Diesel and department stores such as Selfridges in London and Bloomingdales in New York. However, the majority of work being created by designers exists either as self-initiated research projects or in association with arts institutions and funding bodies. Designers are finding that in order to explore the future of physical interfaces further they need to blend commercial practice with a self-initiated commitment to research and development.

For New York-based design group Antenna, working between self-initiated experimental installations and commercial productions has proved to be a positive relationship, as co-founder Sigi Moeslinger explains: 'We get some commercial projects because of the experimental work we have done. In turn, the commercial projects often have to pay for the experimental ones. The experimental projects inform the commercial and vice versa. In both cases we learn things that bring us further with the other.' [17]

Moeslinger and co-founder Masamichi Udagawa combine a background in product design with professional web design experience to create installations that apply interactive thinking to environmental design. Antenna create installations based around the everyday material world 'through appearance, tactility, gestures, what to wear and where things are placed'. [18]

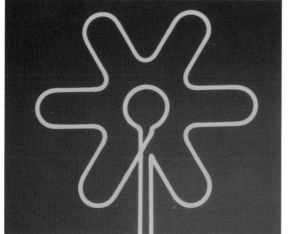

POWER FLOWER
Designers:
Masamichi Udagawa
and Sigi Moeslinger
2002
Photo: Ryuzo Maunaga

By contrasting environmental elements that are familiar and comforting with digital interaction, Antenna hope to invoke new sensations in participants. **THE EMPEROR'S NEW CLOTHES** interactive installation exhibited at Artists Space in New York was designed as a shop display. Visitors were encouraged to pick up a series of coat hangers and carry them into a fitting room. By setting the hanger on a hook the visitor illuminated a 'magic mirror' that would float a series of animated images onto the viewer's reflection, interacting in different ways with the visitor's body and movements.

Antenna experimented with the possibilities of casual participation in a window display for Bloomingdale's in New York entitled **POWER FLOWER**. As people walked along the window display on Lexington Avenue they unknowingly triggered a series of motion sensors, creating a wave-like trail of neon flowers in the wake of every passer by. For Moeslinger, the project existed as a kind of digital self-realization for the people that walked past: 'the injection of an unexpected reaction in an everyday situation may interrupt people in their path, slow them down, and heighten awareness of their environment or even of themselves.' [19]

For Antenna, the move beyond web design to engaging with interactive environments has led to a rediscovery of the joy of experimentation, while at the same time allowing a more tangible connection with their audience: 'We establish a physical exchange between the user and what we design, which is meant to deliver an emotional reaction.' [20]

—→
TYPOGRAPHIC TREE
Design Agency: Digit
2001

DIGITAL AQUARIUM
Design Agency: Digit
Photo: Cathy Vinton
2002
(pp. 116–17)

London-based design agency Digit have successfully cultivated a reputation for award-winning commercial web design, including websites for Habitat, Stella McCartney and MTV2, while balancing their work with a commitment to researching and developing new interactive design methods. Digit have created a series of experimental interactive projects working with arts institutions and commercial sponsors that have developed their research and development ideas.

In 2001 Digit displayed an interactive installation, **TYPOGRAPHIC TREE**, at the ICA in London. This installation created new methods for the public to interact with technology; encouraging visitors to 'grow' screen-based tree elements through sound and allowing individuals to change the design by altering the tone, pitch and volume of their voice. As Daljit Singh, Digit's creative director, explains: 'The initial idea came from looking at how nature could be used to communicate interaction without a keyboard, a mouse or a touch-screen. We wanted to go back to the fundamental ways of communicating…It is important for our projects that the user gets an immediate result – to give the user a level of security and comfort – especially in an exhibition environment.'

Like New York-based designer Joshua Davis, Digit are able to step between the boundaries of art and design practice, and followed their installation at the ICA with another project, **DIGITAL AQUARIUM**, at the Design Museum in London in 2002. Digit worked with Motorola and the Design Museum to create an installation drawing together mobile communications technology and natural forms. Digital Aquarium featured a series of suspended mobile phones. By dialling a choice of numbers printed on the outside of the display case, the user was able to set off a sequence of phone calls within the case – allowing the phones to light up and flicker like shoals of fish, particularly at night.

While Digit have worked with a number of commercial clients, including Nike, Adidas and Motorola, on projects that involve more tangible uses of interactive media, Daljit Singh feels that this area of design is still to be embraced by corporate clients: 'I don't think clients really understand all the possibilities that these technologies can offer, and the driving force behind much of the experimentation going on at the moment is still the interior designer, or the architect or the designer. There is more work still to be done to bring these new kinds of interaction into the mix of mainstream communications techniques, because ultimately clients need to be persuaded that these new kinds of presentation will benefit the brand and provide a tangible service to them.'

FUTURE EXPERIENCES: RETHINKING OUR RELATIONSHIPS

Interacting with technology can sometimes be unsettling, unpredictable and even frightening. At the heart of explorations into the possibilities for physical interfaces and augmented realities lies the desire by designers to create stronger, closer and more tangible connections between the user and technology, and between the user and other people.

Our nervous relationship with technology can be expressed through a fear of loss of control or a fear of technology breaking down, from health concerns over mobile phone masts or the fear of 'network meltdown'. Some designers are choosing to explore this 'interaction anxiety', creating refuges from the complexities of modern living. Designer Toshio Endo has created **WWW.SAFEPLACES.NET**, a temporary respite on the web featuring a sleeping boy, plants and animals and rolling hillsides. As the user interacts with the site, he or she can discover ambient sounds, a hidden passage for links and texts and the gentle and slow regular movement from day to night. For Endo the site exists as 'a rest stop on the web with a simple function…a toy I can give to people. A place you can visit where you don't have to take things so seriously…to chill out, take your time and have fun.' [21]

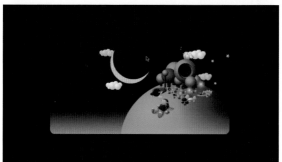

Conceptual designers Anthony Dunne and Fiona Raby see electronic products becoming fully assimilated into our everyday culture and our psyche. These products 'dream', entering our consciousness and forming new kinds of relationships with us. In 2001, Dunne and Raby's **PLACEBO PROJECT** explored the psychological impact of our relationships with electronic products and the electromagnetic radiation they may produce, placing prototype objects in the homes of a group of volunteers that would 'protect' them or 'mediate' between them and the electronic objects in their homes. For Dunne and Raby, these electronic objects develop private lives, hidden from human vision: 'Occasionally we catch a glimpse of this life when objects interfere with each other, or malfunction. Many people believe that mobile phones heat up their ears, or they feel their skin tingle when they sit near a TV, and almost everyone has heard stories of people picking up radio broadcasts in their fillings.' It is of little interest to these designers whether these stories are true or not; instead they are interested in 'the narratives people develop to explain and relate to electronic technologies'. [22]

It is this awareness of the narratives created between users and the technology around them that lies at the heart of designers' investigations into the future of interactive design. This stems from a desire by designers to create greater tactile involvement, and also to produce a means of focusing on the emotional and psychological experiences of the user, whether from a negative or positive perspective. From the enhanced personal networks offered by mobile communications, to the individualized and responsive brand 'experiences' on offer from companies to their consumers, and the new physical environments being created by designers and design students, interactive media is changing not only the way in which we live and relate to each other, but also the way in which we view ourselves.

The relationships between users, technologies and the spaces through which these interactions are transmitted can become, according to writer D.W. Winnicott, the 'third area of human living', a space of creativity and potentiality neither inside nor outside of the world of shared reality. For Winnicott, this space is precisely the location of cultural experience: 'this potential space is at the interplay between there being nothing but me, and there being objects and phenomena outside my omnipotent control'. [23]

In his book 'TechGnosis', the writer Erik Davis describes how technology has come to replace myth and magic in the modern mind: 'Powerful new technologies are magical because they function as magic, opening up novel and protean spaces of possibility within social reality.' [24] In order to realize these magical and transformative possibilities, designers must look beyond the aesthetics of the interactive experience towards an understanding of the cultural, informational, social and even emotional implications of new technologies, for it is at this point that technology and the user can truly come together.

FURTHER RESEARCH

Much information on interactive design can be found in design and new media magazines such as Blueprint, Creative Review, Cre@te Online, Design Week and Wired. There are also a large number of online resources, from art and design portals such as www.designiskinky.net, www.k10k.net, www.rhizome.org and www.threeoh.com to technology information sites such as www.cnet.com, www.geek.com and www.znet.com.

Albrecht, Donald, Ellen Lupton and Steve Skov Holt, Design Culture Now, Laurence King Publishing Ltd, London, 2000.

Austin, Jane (ed.), The Graphics Book, D&AD Masterclass Series, Rotovision, New York, 2002.

Baumgärtel, Tilman, Net.art 2.0: New materials towards Net Art, Institut für moderne Kunst Nürnberg, Nürnberg, 2001.

Bierut, Michael (ed.) et al, Looking Closer, Allworth Press, New York, 1994.

Curran, Steve, Convergence Design, Rockport, Gloucester, Massachusetts, 2003.

Croninger, Curt, Fresh Styles for Web Designers: eye candy from the underground, New Riders Publishing, New York, 2002.

Davis, Erik, Techgnosis: myth, magic and mysticism in the age of information, Serpent's Tail, London, 1999.

Davis, Joshua and Eric Jordan, New Masters of Flash, Friends of ED, 2000.

Dunne, Anthony and Fiona Raby, Design Noir: the secret life of electronic objects, August Media Ltd, London, 2001

Faber, Liz, Re:play: ultimate games graphics, Laurence King Publishing Ltd, London, 1998.

Farrington, Paul, Interactive: the Internet for graphic designers, Rotovision, New York, 2002.

Fiell, Charlotte and Peter (eds), Graphic Design for the 21st Century: 100 of the world's best graphic designers, Taschen, Cologne, 2003.

Foster, Hal, Design and Crime: and other diatribes, Verso, London and New York, 2002.

Futurefarmers (ed.), Harvest: Futurefarmers 1995–2002, Systems Design Ltd, Hong Kong, 2002.

Harkin, James, Mobilisation: the growing public interest in mobile technology, Demos, London, 2003.

Heller, Steven and Marie Finamore (eds), Design Culture: an anthology of writing from the AIGA Journal of Graphic Design, Allworth Press, New York, 1997.

Heskett, John, Toothpicks and Logos: design in everyday life, Oxford University Press, Oxford, 2002.

Kusters, Christian and Emily King, Restart: new systems in graphic design, Thames and Hudson, London, 2001.

Maeda, John, Maeda@Media, Thames and Hudson, London, 2000.

Manovich, Lev, The Language of New Media, MIT Press, Cambridge, Massachusetts, 2001.

Myerson, Jeremy and Graham Vickers, Rewind: 40 years of design and advertising, Phaidon Press Ltd, London and New York, 2002.

Newark, Quentin, What is Graphic Design? Rotovision, New York, 2002.

Odling-Smee, Anne, The New Handmade Graphics: beyond digital design, Rotovision, New York, 2002.

Poyner, Rick, Obey the Giant: life in the image world, August Media Ltd, London, 2001.

Seabrook, John, Nobrow: the culture of marketing, the marketing of culture, Alfred A. Knopf, New York, 2000.

Vartanian, Ivan, Now Loading: the aesthetic of web graphics, Ginko Press, Corte Madera CA, 2001.

NOTES

All quotations in the text, unless otherwise stated, are based on interviews with the author.

Chapter 1
Screen Visions

1. Irene McAra-McWilliam, quoted in Adrian Shaughnessy, 'Life Lessons', Design Week, 27 June 2002, p.14.

2. Yugo Nakamura, quoted in www.designboom.com.

3. Karen Ingram, quoted in www.designiskinky.net.

4. John Maeda, Maeda@Media, Thames and Hudson, London, 2000, p.145.

5. John Maeda, quoted in 'Maeda to Measure', Blueprint, October 2000, p.65.

6. Kevin Robins, Into the Image: culture and politics in the field of vision, Routledge, London and New York, 1996, p.16.

7. Matt Owens, quoted in 'The Power of One', Cre@te Online, July 2001, p.53.

8. Joshua Davis, quoted in www.digital-web.com.

9. Philip Dodd, quoted in Maeve Hosea, 'The fine art of partnership', Design Week, 24 April 2003, p.8.

10. Peter Hall, 'State of Matter', Design Week, 22 May 2003, p.14.

11. Quentin Newark, What is Graphic Design? Rotovision, New York, 2002, p.28.

12. Joachim Blank, quoted in Tilman Baumgärtel, net.art 2.0, Verlag für moderne Kunst, Nürnberg, 2001, p. 123.

13. Dirk Paesmans, quoted in Tilman Baumgärtel, p.173.

14. Brett Stalbaum, 'Conjuring Post-worthlessness: contemporary web art and the postmodern condition', Switch, 1997.

15. Vivian Sobchack, 'New Age Mutant Ninja Hackers: Reading Mondo 2000', The South Atlantic Quarterly, 1993, p.577.

16. Kevin Robins, p.12.

17. Philippe Apeloig, quoted in Charlotte and Peter Fiell (eds), Graphic Design for the 21st Century, Taschen, Cologne, p.70.

18. Matt Owens, quoted in www.youworkforthem.com.

19. Patrick Sundqvist, quoted in www.nervousroom.com.

20. Nik Roope, quoted in Mike Exon, 'On the Up', Design Week The Top 100, June 2003, p.57.

Chapter 2
Styling the Web

1. John Hoyt, quoted in www.designiskinky.net.

2. Steffen Sauerteig, quoted in Francesca Gavin, 'eBoy's toys', Blueprint, July 2002, p.70.

3. Steffen Sauerteig, quoted in Francesca Gavin, p.72.

4. Kai Vermehr, quoted in Francesca Gavin, p.72.

5. Steffen Sauerteig, quoted in Francesca Gavin, p.72.

6. Huw Morgan, quoted in Anne Odling-Smee, The New Handmade Graphics, Rotovision, 2002, p.40.

7. Andrew Blauvelt, 'Towards a complex simplicity', Eye, vol.9, no.35, Spring 2000.

8. Anne Odling-Smee, quoted in Mark Sinclair, 'What touches you is what you touch', Creative Review, March 2003, p.51.

9. Stuart Bailey, quoted in Mark Sinclair, p.51.

10. Alice Rawsthorn, quoted in Sara Manuelli, 'More is More', Design Week, 31 July 2003, p.16.

11. Ben Radatz, quoted in Mark Sinclair, 'The boy from Brazil', Creative Review, December 2002, p.58.

12. Nando Costa, quoted in Mark Sinclair, p.60.

13. Gerard Saint, quoted in Dominic Lutyens, 'Sketch Show', Observer Magazine, 29 June 2003.

14. James McMullan, 'Drawing and Design: an idea whose time has come, again', in Michael Bierut (ed.) et al, Looking Closer, Allworth Press, New York, 1994, p. 171.

15. James Paterson, quoted in www.designiskinky.net.

16. Ibid.

17. Andy Simionato, quoted in Mark Sinclair, 'Desktop publishing', Creative Review, July 2003, p.65.

18. Tibor Kalman, J. Abbott Miller and Karrie Jacobs, Good History/Bad History, in Michael Beirut (ed.) et al, p. 27.

Chapter 3
Branding the Web

1. Mission Statement, www.b-architects.com, June 2003.

2. Mike Yamamoto, 'Browser Revolution: ten years after', www.znet.com, 14 April 2003.

3. Matt Barnard, 'Caution is advised over potential of e-commerce', Design Week, 8 August 2002.

4. Helene Venge, quoted in Hannah Booth, 'Levi's Lateral thinking on website', Design Week, 6 March 2003, p.5.

5. Steve Hayden, 'Ad Space', Wired, June 2003, p.162.

6. Ibid.

7. Martin Raymond, quoted in Clare Dowdy, 'Battle of the Brands', Blueprint, August 2003, p.69.

8. Lisa Strand, quoted in Mike Yamamoto, 'Legacy, a brave new world of the World Wide Web', www.zdnet.com, 14 April 2003.

9. Tammy Savage, quoted in Joe Wilcox, 'Can Microsoft catch the teen spirit?', www.cnetnews.com, 20 February 2003.

10. Wally Olins, 'Yes Logo', The Times Magazine, 13 September 2003, p.86.

11. Ken Frakes, quoted in 'Brands Hatch', Cre@te Online, March 2003.

12. Peter Martin and Daniel Karczinski, 'Branding Interface: Interview with Markus Hanzer', www.icograda.org.

13. 'Which design consultancy does the best work?', Creative Review Peer Poll Results, May 2002, p.26.

14. James Tindall, quoted in Helen Walters, 'Sight and Sound', Creative Review, August 2002, p.34.

15. Florian Schmitt, quoted in Helen Walters, 'Systems Abuse', Creative Review, April 2002, p.66.

16. Phil Knight, quoted in Gavin Lucas, 'Which brand has the best advertising?', Creative Review Peer Poll 2003, September 2003, p.56.

17. Nick Barham, quoted in Paula Carson, 'Let me entertain you', Creative Review, December 2002, p.56.

18. Habbo Company Profile, www.habbogroup.com, January 2003.

19. Nokia Game website, www.nokiagame.com, September 2003.

20. John Seabrook, Nobrow: the culture of marketing, the marketing of culture, Alfred A. Knopf, New York, 2000.

21. Rick Poyner, Obey the Giant: life in the image world, August Media Ltd, London, 2001.

Chapter 4
Future Interactions

1. Irene McAra-McWilliam, quoted in Adrian Shaughnessy, 'Life Lessons', Design Week, 27 June 2002, p.14.

2. Frank Rose, 'Europeans do it better', Wired, August 2003, p.25.

3. Robyn Greenspan, 'Broadband's reach gets broader', www.cyberatlas.com, 7 February 2003.

4. James Penhune, quoted in Robyn Greenspan, 7 February 2003.

5. Henley Management College, 'Young People could not live without their mobiles', Press Association, 12 May 2003.

6. James Harkin, Mobilisation: the growing public interest in mobile technology, Demos, London, June 2003.

7. Mori Omnibus Survey 2001, quoted in James Harkin, June 2003.

8. Kate Fox, Evolution, Alienation and Gossip: the role of telecommunications in the 21st century, Social Issues Research Centre, Oxford, 2001.

9. James Harkin, June 2003.

10. Frazer Jelleyman, quoted in 'TBWA put the fun back into PS2', Creative Review, August 2003.

11. Tota Hasegawa, quoted in Creative Review, May 2002, p.12.

12. Walter Deffor, quoted in 'World's first GPRS game launched', www.hi2.com, 30 August 2001.

13. Steve Curran, Convergence Design, Rockport, Gloucester, Massachussetts, 2003, p.123.

14. 'Congestion Charging: summary of week 6', Transport for London press release, www.londontransport.co.uk, 1 April 2003.

15. James Harkin, June 2003.

16. Drew Leder, The Absent Body, University of Chicago Press, Chicago, 1990, p.118.

17. Sigi Moeslinger, quoted in Neil Churcher, 'Aerialism', Design Week, 13 February 2003, p.17.

18. Ibid.

19. Ibid.

20. Masamichi Udagawa, quoted in Neil Churcher, p.16.

21. Toshio Endo, quoted in Ivan Vartanian, Now Loading: the aesthetic of web graphics, Ginko Press, Corte Madera CA, p.34.

22. Anthony Dunne and Fiona Raby, Design Noir: the secret life of electronic objects, August Media Ltd, London, 2001, p.75.

23. D.W. Winnicott, Playing with reality, Penguin, London, 1982, p.118.

24. Erik Davis, Techgnosis: myth, magic and mysticism in the age of information, Serpent's Tail, London, 1999.